GUIDELINES FOR

Use of Vapor Cloud
Dispersion Models

Publications Available from the
CENTER FOR CHEMICAL PROCESS SAFETY
of the
AMERICAN INSTITUTE OF CHEMICAL ENGINEERS

Guidelines for Use of Vapor Cloud Dispersion Models, Second Edition
Guidelines for Evaluating Process Plant Buildings for External Explosions and Fires
Guidelines for Chemical Transportation Risk Analysis
Guidelines for Safe Storage and Handling of Reactive Materials
Guidelines for Technical Planning for On-Site Emergencies
Guidelines for Process Safety Documentation
Guidelines for Safe Process Operations and Maintenance
Guidelines for Process Safety Fundamentals in General Plant Operations
Guidelines for Chemical Reactivity Evaluation and Application to Process Design
Tools for Making Acute Risk Decisions with Chemical Process Safety Applications
Guidelines for Preventing Human Error in Process Safety
Guidelines for Evaluating the Characteristics of Vapor Cloud Explosions, Flash Fires, and BLEVEs
Guidelines for Implementing Process Safety Management Systems
Guidelines for Safe Automation of Chemical Processes
Guidelines for Engineering Design for Process Safety
Guidelines for Auditing Process Safety Management Systems
Guidelines for Investigating Chemical Process Incidents
Guidelines for Hazard Evaluation Procedures, Second Edition with Worked Examples
Plant Guidelines for Technical Management of Chemical Process Safety, Rev. Ed.
Guidelines for Technical Management of Chemical Process Safety
Guidelines for Chemical Process Quantitative Risk Analysis
Guidelines for Process Equipment Reliability Data, with Data Tables
Guidelines for Vapor Release Mitigation
Guidelines for Safe Storage and Handling of High Toxic Hazard Materials
Understanding Atmospheric Dispersion of Accidental Releases
Expert Systems in Process Safety
Concentration Fluctuations and Averaging Time in Vapor Clouds
Safety, Health, and Loss Prevention in Chemical Processes: Problems for Undergraduate Engineering Curricula
Safety, Health, and Loss Prevention in Chemical Processes: Problems for Undergraduate Engineering Curricula—Instructor's Guide
Proceedings of the International Conference and Workshop on Modeling and Mitigating the Consequences of Accidental Releases of Hazardous Materials, 1995.
Proceedings of the International Symposium and Workshop on Safe Chemical Process Automation, 1994
Proceedings of the International Process Safety Management Conference and Workshop, 1993
Proceedings of the International Conference on Hazard Identification and Risk Analysis, Human Factors, and Human Reliability in Process Safety, 1992
Proceedings of the International Conference/Workshop on Modeling and Mitigating the Consequences of Accidental Releases of Hazardous Materials, 1991.
Proceedings of the International Symposium on Runaway Reactions, 1989
CCPS/AIChE Directory of Chemical Process Safety Services

GUIDELINES FOR

Use of Vapor Cloud Dispersion Models

SECOND EDITION

CENTER FOR CHEMICAL PROCESS SAFETY

of the

American Institute of Chemical Engineers
345 East 47th Street, New York, NY 10017

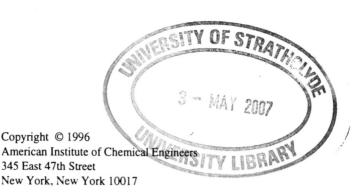

Library of Congress Cataloging-in Publication Data
Guidelines for use of vapor cloud dispersion models / Center for
    Chemical Process Safety of the American Institute of Chemical
    Engineers.
        p.   cm.
    Includes bibliographical references and index.
    ISBN 0–8169–0702–1
        1. Atmospheric diffusion—Mathematical models.   2. Hazardous
    substances—Environmental aspects—Mathematical models.
    3. Vapor clouds—Mathematical models.   I. American Institute of
    Chemical Engineers. Center for Chemical Process Safety.
    QC880.4.D44G85   1996
    628.5′3′0113—dc20   96–26950

It is sincerely hoped that the information presented in this document will lead to an even more
impressive safety record for the entire industry: however, the American Institute of Chemical
Engineers, its consultants. CCPS subcommittee members, their employers, their employers'
officers and directors and EARTH TECH disclaim making or giving any warranties or
representations, express or implied, including with respect to fitness, intended purpose, use or
merchantability and/or correctness or accuracy of the content of the information presented in
this document. As between (1) the American Institute of Chemical Engineers, its consultants.
CCPS subcommittee members, their employers, their employers' officers and directors and
EARTH TECH and (2) the user of this document, the user accepts any legal liability or
responsibility whatsoever for the consequence of its use or misuse.

Contents

Preface

For 40 years the American Institute of Chemical Engineers (AIChE) has been involved with process safety and loss control issues in the chemical, petrochemical, hydrocarbon process, and related industries and facilities. AIChE publications and symposia are information resources for the chemical engineering and other professions on the causes of process incidents and the means of preventing their occurrences and mitigating their consequences.

The Center for Chemical Process Safety (CCPS), a directorate of AIChE, was established in 1985 to develop and disseminate technical information for use in the prevention of major chemical process incidents. With the support and direction of the CCPS Advisory and Managing Boards, a multifaceted program was established to address the need for process safety management systems to reduce potential exposures to the public, facilities, personnel, and the environment. This program involves the development and publication of guidelines related to specific areas of process safety management; organizing convening, and conducting seminars, symposia, training programs, and meetings on process safety-related matters; and cooperation with other organizations, both internationally and domestically, to promote process safety. CCPS's activities are supported by funding and expertise from over 90 entities.

In 1987 CCPS published *Guidelines for Use of Vapor Cloud Dispersion Models*, and in 1989, *Workbook of Test Cases for Vapor Cloud Source Dispersion Models*. These books have served well but are now outdated. At nearly a decade old, they refer to an earlier generation of vapor cloud models.

The present book has been expanded to include both source term models and vapor cloud dispersion models, and it incorporates worked examples with the model descriptions.

Acknowledgments

The American Institute of Chemical Engineers and the Center for Chemical Process Safety (CCPS) express their gratitude to all the members of the Vapor Cloud Modeling Subcommittee for their unstinting efforts and technical contributions in the preparation of this *Guidelines*. The members of this distinguished group are:

Ronald J. Lantzy, Chair   *Rohm and Haas Company*
Gib R. Jersey, Vice Chair   *Mobil Technology Company*
William J. Hague, Past Chair *AlliedSignal, Inc.*
Douglas N. Blewitt   *Amoco Corporation*
Sanford G. Bloom   *Lockheed Martin Energy Systems*
Donald J. Connolley   *AKZO Nobel Chemicals, Inc.*
George E. DeVaull   *Shell Oil Company*
Seshu Dharmavaram   *DuPont Company*
Ebrahim Esmaili   *Exxon Research and Engineering Co.*
David J. Fontaine   *Chevron Research and Technology Co.*
Gene K. Lee   *Air Products and Chemicals, Inc.*
John T. Marshall   *Dow USA, Texas Operations*
David McCready   *Union Carbide Corporation*
Robert Moser   *Cigna Property and Casualty*
Ronald D. Myers   *Rohm and Haas Company*
Malcolm L. Preston   *ICI Engineering Technology*
Jerry M. Schroy   *Monsanto Company*
Kenneth W. Steinberg   *Exxon Research and Engineering Co.*
Jawad Touma   *U.S. Environmental Protection Agency*

CCPS guidance and counsel were appropriately provided by its recent director, Bob G. Perry, and its current director, Jack Weaver. Liaison between the Subcommittee and CCPS was provided by William J. Minges, CCPS Staff.

The contract for preparing the *Guidelines* was awarded to EARTH TECH (formerly Sigma Research Corp.), Concord, MA. The following people are the authors of this book:

> Steven R. Hanna, EARTH TECH.
> Peter J. Drivas, Gradient Corp.
> Joseph J. Chang, EARTH TECH.

We acknowledge the dozens of scientists and engineers who prepared useful discussions of their models, who took the time to complete questionnaires, and who sent copies of relevant manuscripts to EARTH TECH.

CCPS also expresses its appreciation to members of the Technical Steering Committee for their valuable advice and support.

The group to whom the Subcommittee is especially indebted consists of those who volunteered to provide peer review:

J. Steven Arendt	*JBF Associates, Inc.*
Daniel A. Crowl	*Michigan Technological University*
Thomas O. Gibson	*Dow Chemical Company*
Kenneth Harrington	*Battelle*
Michael J. Hitchler	*Westinghouse Savannah River Company*
John A. Hoffmeister	*Lockheed Martin Energy Systems*
John Hudson	*PCR, Inc.*
Dimitrios Karydas	*Factory Mutual Research Corporation*
Steven Kent	*Raytheon Engineers and Constructors, Inc.*
Georges A. Melham	*Arthur D. Little, Inc.*
Kenneth Mosig	*AIU Energy/Starr Technical Risks Agency, Inc.*
John A Noronha	*Eastman Kodak Company*
Frank P. Ragonese	*Mobil Oil Corporation*

Nomenclature

$\overline{C}$ (kg m^{-3})	Mean or average concentration	6.3	
C_0' (kg m^{-3})	Observed concentration random variability	6.6	(6-18)
C_p' (kg m^{-3})	Predicted concentration random variability	6.6	(6-19)
C'' (kg kg^{-1})	Concentration of gas in plume	5.6	(5-51)
C_l'' (kg kg^{-1})	Concentration of liquid in plume	5.6	(5-53)
C_{cl} (kg m^{-3})	Concentration on plume centerline	6.3	
C_I	Cloudiness index	4.2.5	(4-25)
C_m (kg m^{-3})	Mean concentration	5.5	
C_i (kg m^{-3})	Initial concentration	5.2.2	(5-13)
C_O (kg m^{-3})	Observed concentration	6.6	(6-18)
C_p (kg m^{-3})	Predicted concentration	6.6	(6-18)
C_{peak} (kg m^{-3})	Peak concentration	6.3	
d_t (m)	Tank diameter	4.2.2	(4-11)
D (m^{2}s^{-1})	Molecular diffusivity of pollutant gas	5.9.2	(5-75b)
D_B (m^{2}s^{-1})	Brownian diffusivity for particles	5.9.2	(5-76)
D_c (m)	Source dimension for continuous release	5.5	
D_i (m)	Source dimension for instantaneous release	5.5	
D_o (m)	Initial cloud width	5.1	(5-1)
D_p (m)	Pipe diameter	4.2.1	(4-6)
D_p (μm)	Particle diameter	5.9.1	(5-67)
ERPG	Emergency Response Planning Guidelines	3.4	
f	Pipe friction factor	4.2.1	(4-6)
f_{area}	Area view factor	4.2.5	(4-24)
F_{Dp} (kg m^{-2}s^{-1})	Particle deposition flux	5.9.1	(5-69)
F_{wet} (kg m^{-2}s^{-1})	Wet deposition flux	5.9.2	(5-80)
$FAC2$	Fraction within a factor of two	8.2	(8-4)
FB	Fractional bias	8.2	(8-5)
g (ms^{-2})	Acceleration of gravity (9.8 ms^{-2})	3.3	(3-1)
g_o (ms^{-2})	Reduced gravity = $g(\rho_o - \rho_a)/\rho_a$	5.5	
G (kg s^{-1}m^{-2})	Mass emission rate per unit area	4.2.3.2	(4-15)
h_c (Wm^{-2} K^{-1})	Heat transfer constant	5.3.3	(5-38)
h_o (m)	Initial cloud depth	5.3.2	

h_p (m)	Plume centerline height above ground	5.4	(5-40)
h_s (m)	Stack height	5.1	(5-4a)
H_B (m)	Building height	5.7.1	(5-62)
H_f (m K s^{-1})	Sensible heat flux	3.3	(3-1)
H_l (m)	Height of liquid above puncture	4.2.2	(4-10)
H_s (W m^{-2})	Convective heat flux	5.3.3	(5-38)
H_{vap} (J kg^{-1})	Heat of vaporization of liquid	4.2.3.1	(4-14)
H_{WC} (m)	Plume depth, worst case	2.1	(2-1)
I	Intermittency of concentration record	6.4	(6-13)
k_g (m s^{-1})	Mass transfer coefficient	4.2.5.2	(4-29)
k_s (J s^{-1} m^{-1}K^{-1})	Thermal conductivity of soil	4.2.5.1	(4-28)
K_F, K_A, K_g	Dimensionless factors	4.2.1	(4-6)
K (m^2 s^{-1})	Eddy diffusivity vector	5.6	(5-49)
L (m)	Monin-Obukhov length	3.3	(3-1)
L_c (m)	Length of building wake or cavity	5.7.4	
L_p (m)	Length of pipe	4.2.1	(4-6)
L_t (m)	Length of tank	4.2.2	(4-13)
LAI	Leaf Area Index	5.9.2	(5-78b)
m (kg)	Liquid mass in pool	4.2.5	(4-22)
m_j (kg)	Mass of gas component j	5.3.2	(5-17)
m_t (kg)	Total mass in pipeline	4.2.1	(4-5)
M (kg kg-mole^{-1})	Molecular weight	4.2.1	
M_j (kg kg-mole^{-1})	Molecular weight of gas component j	5.3.2	(5-17)
M_o (m^4 s^{-2})	Initial plume momentum flux	5.2.1	(5-17)
MG	Geometric mean	8.2	(8-1)
n_{To}	Total moles of liquid	4.2.6	(4-34)
N_{Re}	Reynolds number	4.2.5.2	(4-30)
N_{Sc}	Schmidt number	4.2.5.2	(4-30)
N_{Sh}	Sherwood number	4.2.5.2	(4-30)
N_{St}	Stokes number	5.9.2	(5-77)
p	Probability density function	6.3	(6-6)
p (N m^{-2})	Tank pressure	4.25.1	(4-1)

p_a (N m^{-2})	Ambient pressure	4.2.1	(4-1)
p_{is} (N m^{-2})	Saturation vapor pressure of component i	4.2.6	(4-34)
p_o (N m^{-2})	Stagnation pressure	4.2.3.2	(4-15)
p_v (N m^{-2})	Vapor pressure	4.2.3.2	(4-15)
P (mm hr^{-1})	Precipitation rate	5.9.2	(5-81)
P	Cumulative density function	6.3	(6-7)
q_a (kg s^{-1} m^{-2})	Evaporation rate per unit area	4.2.6	(4-34)
q_{cond} (J s^{-1})	Conduction heat transfer from ground	4.2.5	(4-22)
q_{conv} (J s^{-1})	Convection heat transfer from air	4.2.5	(4-22)
q_{evap} (J s^{-1})	Heat loss due to evaporation	4.2.5	(4-22)
q_{rad} (J s^{-1})	Radiative heat transfer from air	4.2.5	(4-22)
q_{sun} (J s^{-1})	Incident solar radiation	4.2.5	(4-22)
Q (kg s^{-1})	Source mass emission rate	2.1	(2-1)
Q_a (kg s^{-1})	Evaporation rate of liquid pool	4.2.5	(4-26)
Q_f (kg s^{-1})	Mass emission rate of liquid that flashes	4.2.3.1	(4-14)
Q_i (kg)	Mass in instantaneous cloud release	5.4	(5-4)
Q_l (kg s^{-1})	Liquid mass flow rate	4.2.2	(4-10)
Q_o (kg s^{-1})	Initial source mass emission rate	4.2.1	(4-4)
r (m)	Distance on building from source to receptor	5.7.2	(5-64)
r_a (s m^{-1})	Aerodynamic resistance	5.9.2	(5-70)
r_{cut} (s m^{-1})	Cuticle resistance	5.9.2	(5-78b)
r_f (s m^{-1})	Stomate resistance	5.9.2	(5-78b)
r_g (s m^{-1})	Resistance to transfer across surface	5.9.2	(5-78b)
r_l (m)	Liquid pool radius	4.2.4	(4-18)
r_s (s m^{-1})	Surface resistance	5.9.2	(5-70)
r_t (s m^{-1})	Transfer resistance	5.9.2	(5-70)
R	Correlation coefficient	8.2	(8-3)
R (m)	Plume radius	5.2.2	(5-7)
R (J K^{-1} kg^{-1})	Gas constant for specific gas	5.3.2	(5-17)
R^*	Universal gas constant (8310 J K^{-1} kg-mole^{-1})		4.2.1
Ri_*	Local cloud Richardson number	5.3.2	(5-30)
Ri_a	Ambient Richardson number	5.6	(5-60)
Ri_o	Critical Richardson number	5.1	(5-1)

s (m)	Distance along plume axis	5.2.2	(5-7)
S_{CF}	Slip correction factor	5.9.1	(5-68)
t (s)	Time	4.2.1	(4-4)
t_e (s)	Exposure time	6.5	
T' (K)	Temperature turbulent fluctuation	3.3	(3-3)
T_{air} (K)	Temperature, air	3.1	
T_b (K)	Normal boiling point of liquid	4.2.3.1	(4-14)
T_d (s)	Time duration of release	5.5	
T_I (s)	Integral time scale	6.3	(6-9)
T_o (K)	Temperature, tank	4.2.1	
T_p (K)	Pool temperature	4.2.5	(4-22)
T_s (s)	Sampling time	6.3	
T_{soil} (K)	Soil temperature	4.2.5.1	(4-28)
u (m s^{-1})	Wind speed	3.3	(3-4)
u' (m s^{-1})	Wind speed turbulent fluctuation	3.3	(3-2)
$\mathbf{u}$ (m s^{-1})	Wind vector	5.6	(5-49)
u_1 (m s^{-1})	Wind speed at height of 1 m	9.1	
u_* (m s^{-1})	Friction velocity	3.3	(3-1)
u_a (m s^{-1})	Advection velocity of cloud	5.3.2	(5-20)
u_e (m s^{-1})	Entrainment rate	5.2.1	
u_H (m s^{-1})	Wind speed at height of building	5.7.2	(5-63)
u_s (m s^{-1})	Plume speed along axis	5.2.2	(5-7)
u_{st} (m s^{-1})	Wind speed at stack height	E	(E-9)
u_{wc} (m s^{-1})	Wind speed, worst case	2.1	(2-1)
v (m s^{-1})	Gross entrainment velocity	5.3.2	(5-26)
v_d (m s^{-1})	Dry deposition velocity	5.9.2	(5-70)
v_{edge} (m s^{-1})	Edge entrainment velocity	5.3.2	(5-28a)
v_{lg} (m^3 kg^{-1})	Difference in specific volume between liquid and gas	4.2.3.2	(4-15)
v_s (m s^{-1})	Gravitational settling speed	5.9.1	(5-67)
v_{top} (m s^{-1})	Top entrainment velocity	5.3.2	(5-28a)
V_c (m^3 s^{-1})	Volume flow rate	5.3.2	(5-28a)
V_{co} (m^3 s^{-1})	Initial volume flow rate	5.1	(5-1)

V_i (m^3)	Volume of instantaneous cloud	5.3.2	(5-18)
V_{io} (m^3)	Initial volume of instantaneous cloud	5.1	(5-2)
V_l (m^3)	Volume of liquid in tank	4.2.2	(4-11)
V_o (m^3)	Volume of instantaneous spill	4.2.4	(4-18)
VG	Geometric variance	8.2	(8-2)
w (m s^{-1})	Local jet vertical speed	5.2.1	
w' (m s^{-1})	Vertical wind speed fluctuation	3.3	(3-3)
w_o (m s^{-1})	Initial jet velocity	5.2.1	
W (m)	Plume width	8.4	
W_B (m)	Width of building	5.7.2	
W_c (m)	Width of canyon between buildings	5.7.1	(5-61)
W_{wc} (m)	Plume width, worst case	2.1	(2-1)
x (m)	Distance from cloud to receptor	5.1	(5-2)
x_g (m)	Distance where plume touches ground	5.2.2	(5-12)
x_i	Initial liquid mole fraction of component i	4.2.6	(4-34)
x_o (m)	Alongwind position of cloud center	5.4	(5-41)
x_v (m)	Virtual source distance	5.4	
y (m)	Lateral position	5.4	(5-40)
y_o (m)	Lateral centerline position of plume	5.4	(5-40)
z (m)	Height above ground	3.3	(3-4)
z_d (m)	Reference height (usually 10 m)	5.9.2	(5-71)
z_o (m)	Roughness length	3.2	
z_p (m)	Plume height above source	5.2.1	(5-5)
α	Mass conservation factor	4.2.1	(4-4)
α_s (m^2 s^{-1})	Thermal diffusivity of soil	4.2.5.1	(4-28)
β (s)	Time constant for release from pipe	4.2.1	(4-4)
$\gamma = c_p/c_v$	Gas specific heat ratio	4.2.1	(4-1)
δ	Dirac delta function	6.4	(6-13)
ΔC_o (kg m^{-3})	Data error in observed concentration	6.6	(6-18)
ΔC_p (kg m^{-3})	Error in predicted concentration due to data errors	6.6	(6-19)

Δh (m)	Maximum rise of dense plume	5.2.2	(5-11)
Δx_{upwind} (m)	Distance dense cloud travels upwind	5.3.2	(5-22)
Δy (m)	Crosswind distance traveled by plume	5.2.1	(5-6)
$\Delta y_{initial}$ (m)	Distance cloud spreads laterally at source	5.3.2	(5-23)
∇	Del operator	5.6	(5-49)
ε	Liquid emissivity	4.2.5	(4-24)
ε_d ($m^2\,s^{-3}$)	Eddy dissipation rate	5.2.2	(5-15)
Λ (s^{-1})	Wet removal inverse time scale	5.9.2	(5-79)
μ ($kg\,m^{-1}\,s^{-1}$)	Air viscosity	5.9.1	(5-67)
v($m^2\,s^{-1}$)	Molecular viscosity of air	5.9.2	(5-75a)
ϕ_s (°)	Solar angle above horizon	4.2.5	(4-25)
ϕ	Dimensionless function of Ri	5.6	(5-57)
ρ_a ($kg\,m^{-3}$)	Ambient air density	5.1	(5-1)
ρ_{aer} ($kg\,m^{-3}$)	Mass of aerosol per unit cloud volume	5.3.2	(5-16)
ρ_l ($kg\,m^{-3}$)	Liquid density	4.2.2	(4-11)
ρ_o ($kg\,m^{-3}$)	Gas density in tank	4.2.1	(4-2)
ρ_{po} ($kg\,m^{-3}$)	Initial plume density	5.1	(5-1)
ρ_s ($kg\,m^{-3}$)	Density of underlying surface	5.3.3	(5-37)
ρ_{sp} ($kg\,m^{-3}$)	Solid particle density	5.9.1	(5-67)
ρ_w ($kg\,m^{-3}$)	Density of water	4.2.4	(4-20)
σ_c ($kg\,m^{-3}$)	Standard deviation of concentration fluctuations	6.4	(6-15)
$\sigma_{SB} = 5.67 \times 10^{-8}\,J\,s^{-1}m^{-2}K^{-4}$	Stefan-Boltzmann constant	4.2.5	(4-24)
σ_x (m)	Alongwind dispersion parameter	5.4	(5-41)
σ_y (m)	Lateral dispersion parameter	5.3.1	
σ_{yl} (m)	Lateral dispersion of instantaneous puff	6.3	(6-3)
σ_z (m)	Vertical dispersion parameter	5.3.1	
θ (°)	Angle of plume axis with horizontal	5.2.2	(5-7)
θ_l (°)	Angle of liquid surface in tank	4.2.2	(4-13)

1

Background and Objectives

The American Institute of Chemical Engineers (AIChE) has a long history of involvement with process safety and loss control for chemical and petrochemical plants. Early in 1985 the AIChE established the Center for Chemical Process Safety (CCPS) to serve as a focus for a continuing program for process safety. The CCPS has published several *Guideline* books on specific technical subjects, including the first edition of the current book, written by S. R. Hanna and P. J. Drivas in 1987, and a related workbook of example model applications written by S. R. Hanna and D. G. Strimaitis in 1989. The decision was made to revise and combine these two documents and update them to reflect advances over the past several years in the field of vapor cloud dispersion modeling. This has been an active field of research and applications, and much new information has been gained since the first edition was written.

The modeling of accidental releases of hazardous or toxic materials is a rapidly evolving field in chemical engineering, driven by the efforts of the industry to prevent and mitigate such incidents, as well as by public concern beginning to be reflected in regulatory requirements. In most cases the word "model" in this document is synonymous with the words "computer program." Overall, the objective of the *Guidelines for Use of Vapor Cloud Dispersion Models*, Second Edition, is to help facilitate the development and use of dispersion modeling as an everyday tool within the industry, along with an understanding of the limitations of that tool. The second edition of the *Guidelines* is intended to provide in one publication an overview of the subject that will be of value to the beginner, the expert, and the manager—whether in industry, in a government agency, or in a university. For the beginner, it is a starting point that will provide perspective and facilitate rapid progress into the field of dispersion modeling. For experts, it is an overview document and a resource, perhaps useful as a text for those apprenticing under their care. For managers, it is a bridge to the expert, permitting the manager to fully understand the difficulties, uncertainties, and limitations associated with modeling accidental releases.

The objectives of this *Guidelines* book are to provide the reader with the following information:

- A practical understanding of the basic physical and chemical principles.
- Guidance on how to select release scenarios.
- Information to permit the modeler to choose the best available model (computer program) for the situation, and knowledge of the expected uncertainty of the model.
- Information on how to run some models and interpret the model outputs.

This document purposely presents the material listed above in a concise way so that it can be easily used. An extensive reference list is provided for readers who wish to explore specific topics in more depth.

The description of twenty-two practical vapor cloud models or computer programs in Section 7 is based on the results of questionnaires sent to model developers. Rather than rely on third-person reviews of models, the developers were requested to describe the capabilities of their models. This list is thought to include all of the more commonly used models, as of about April 1995. In the future, many of these models can be expected to be revised and new models will be developed as our knowledge increases.

Seven examples of typical accidental chemical release scenarios are described and commonly used vapor cloud models are applied to the scenarios in Section 9 and in the Appendices to this *Guidelines* book. These exercises are included in order to give readers an idea of how to select an appropriate model, set up model input files, run the models, and interpret the results. A diskette containing the model input and output files for the seven scenarios is included in a pocket inside the back cover of this book.

The AIChE/CCPS has published a related document, written by DeVaull et al. (1995), and titled *Understanding Atmospheric Dispersion of Accidental Releases*. That document is much shorter than the *Guidelines* and is intended to be more of a "primer" for persons encountering this topic for the first time. The current volume provides more details concerning the technical rationale for the models and includes the section in which seven worked examples are described.

2

Overview of Modeling Procedures, Including Rationale for Selecting Scenarios for Worked Examples

2.1. Types of Scenarios and Models

A few examples of catastrophic releases of hazardous chemicals have been heavily reported by the news media, but a more likely release is much smaller. It is stressed that the same vapor cloud models (i.e., computer programs) are used to treat the large, medium, and small release scenarios, since the basic physical principles are independent of the size of the release. Typical source scenarios can be grouped into three broad areas:

(1) Liquid spill and subsequent evaporation.
(2) Two-phase pressurized release.
(3) Gas release.

There are a variety of environmental impact questions that must be taken into account, related to the location, magnitude, and duration of a certain air concentration. There may be toxicological and flammability information that would strongly influence the types of impacts and averaging times to be modeled. The engineer should carefully check these data for the chemical in question before proceeding with any model calculations.

The choice of a source and dispersion model is dependent on the chemical, the release scenario, and the desired averaging time. Chapters 4 and 5 provide detailed summaries of the physical principles underlying the source release and dispersion models, respectively. The major regimes of hazardous gas modeling in the atmosphere are drawn in Figures 2-1a and 2-1b. Figure 2.1a, for a jet, shows how the initial source emissions and acceleration phases give way to a regime in which the internal buoyancy (positive or negative) of the plume or puff dominates the dispersion. For some strong jets, the jet itself can generate so much entrainment of ambient air that the plume may be diluted such that dense gas effects are not longer important. The dense gas regime is always followed by a transition to a regime in which ambient turbulence dominates the dispersion. Figure 2.1b, for a ground- level area source, shows how the modeling begins with the regime in which ambient density dominates. Field experiments with typical dense gas clouds show that the transition to dispersion dominated by ambient turbulence can occur within 10 m of the source for small release rates (e.g., 0.1 kg/s or less) but can be delayed to distances of 1000 m or greater for

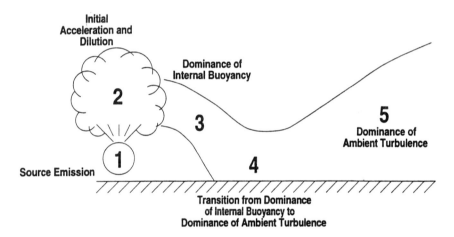

Figure 2-1a. Illustration of five major regimes of hazardous gas modeling for a jet source whose density is initially greater than that of the ambient air.

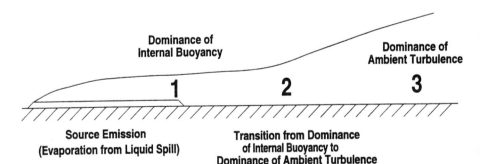

Figure 2-1b. Illustration of three major regimes of hazardous gas modeling for a dense cloud resulting from evaporation from a liquid spill.

large release rates (e.g., 10 kg/s or greater). For some questions, such as the lower flammability limit (LFL) of LNG, the calculations need not go beyond the gravity slumping (dominance of internal buoyancy) regime, since the LFL occurs at a relatively high concentration of a few percent. However, for questions such as the ten-minute average concentration of a toxic gas at a hospital located 10 km from the release, the calculations will probably extend well into the regime where ambient turbulence is dominant, since the health effects of the gas may be important at concentrations as low as 0.1 to 1.0 ppm. Information on concentrations related to health effects for various chemicals is given by, for example, the AIHA (1995) *Emergency Response Planning Guidelines* or the NIOSH (1995) *Pocket Guide to Chemical Hazards.*

In modeling vapor cloud dispersion, there are a number of steps that must be followed, with many pathways and options. A figure containing all the steps and options would be much too long and confusing to present here. Figure 2-2 is a one-page summary of the general sequence of logic or model steps that should be followed. It also serves as an outline for the remainder of the discussions in this book. It is stressed that there are differences in the capabilities of available modeling systems to account for the various effects in Figure 2-2. For example, most vapor cloud dispersion models treat entrainment into dense gas clouds while very few models treat chemical reactions in vapor clouds. Also, compromises often have to be made in the specification of the required input data. The figure shows that the model outputs may be used in risk-management decisions, which are not the subject of this book, but which are discussed in other AIChE/CCPS books, such as the *Guidelines for Chemical Process Quantitative Risk Analysis* (AIChE/CCPS, 1989).

In any particular scenario, only the relevant parts of the sequence in Figure 2-2 would be followed. For example, for a high-velocity CO_2 gas jet emitted directly upward, it is not necessary to know the environmental properties of the underlying surface and only the gas jet source emission model need be used. In this example, the CO_2 concentrations would be reduced to safe levels due to entrainment into the jet before it could interact with the near-ground boundary layer.

The Environmental Protection Agency (EPA) makes available a set of air quality models for calculating the environmental effects of routine air pollutant releases (e.g., SO_2 released from a power plant stack, which would be modeled with the Industrial Source Complex (ISC) model). Touma et al. (1995b) provide an overview of these regulatory models, describe methods for updating the models, and include extensive reference lists. A

SCENARIO DESCRIPTION

Liquid Spill   Pressurized Two-Phase Jet   Vapor Release

DEFINITION OF INPUT PARAMETERS

*Chemical and Physical
Properties of Release*   *Receptor Properties*   *Environmental Properties*

Time duration, location   Locations of receptors   Air temperature, wind speed,
                                                                humidity, stability, mixing
Temperature, pressure,   Averaging time   height, solar radiation
    volume, mass, molecular
    weight, specific and latent   Concentrations of   Soil or water temperature,
    heat, boiling point, thermal   interest   water content, conductivity,
    conductivity   porosity

Geometry of hole or spill   Surface roughness
    (height, diameter, direction)
                                                                Terrain slope and obstructions
Nearby building and storage
    tank dimensions

SOURCE EMISSION MODEL

Catastrophic   Evaporation   Liquid jet   Two-phase jet   Gas jet   Rain out
vessel rupture   from spill

In all cases, variation of emission with time is given

VAPOR CLOUD DISPERSION MODEL

Entrainment   *Thermodynamics*   *Removal*   *Special Effects*

Jet   Heat exchanges with   Chemical reactions   Concentration
Dense gas   environment   Dry deposition   fluctuations
Buoynat gas   Condensation and   Wet deposition   Averaging time
Neutral gas   evaporation   Building downwash
Passive   Heat exchanges due to   In-building
                       chemical reactions   concentrations
                   Lift-off

CONCENTRATION PREDICTION

Output of model—tables or graphs of concentrations for
prescribed locations, heights, and averaging times

Figure 2-2. Generalized vapor cloud model logic sequence. In any specific case, only a
portion of the sequence would be followed.

model such as ISC is easy to run and is recognized by regulatory agencies. However, none of the models discussed by Touma et al. (1995b) apply to dense gas releases. Even though many vapor cloud models have been developed for accidental releases of dense gases, none have been formally "approved" by the EPA. Models such as DEGADIS, TSCREEN, and SLAB are made available to users on the EPA electronic bulletin board. Table 2-1 contains a comparison of some of the major characteristics of the two types of models (i.e., routine models such as ISC versus accidental release models such as HGSYSTEM). Before choosing a model, a user should first consider whether the release scenario could possibly fall under the "routine" category (i.e., continuous releases of neutrally or positively buoyant gases from a stack). Criteria are given in Section 5 (see equations 5-1 through 5-4) for deciding whether dense gas effects are likely to be important for a given scenario.

2.2. Gross Screening Analysis

It is wise to carry out a gross screening analysis prior to embarking on a modeling exercise, since it may be that the predicted concentrations are orders of magnitude below some limits. This screening analysis can also be used to check the magnitudes of the results of subsequent detailed model applications. The following formula can be used for roughly calculating concentrations resulting from emissions near ground-level:

$$C = \frac{Q}{u_{wc}H_{wc}W_{wc}} \qquad (2\text{-}1)$$

where C is concentration (kg/m^3), Q is the source emission rate (kg/s), u_{wc} is the "worst case" wind speed (assumed to equal 1 m/s), H_{wc} is the worst case cloud depth (assumed to equal 50 m), and W_{wc} is the worst case cloud width (assumed to equal $0.1x$, where x is distance from the source). The source term, Q, is assumed to be the mass released during a ten-minute period.

As an example of the use of equation (2-1), suppose that 0.1 kg of ammonia were released accidentally over ten minutes, and that the concern is whether concentrations were high at a day care center 10,000 m from the source. Then, assuming the cloud depth is 50 m and the cloud width is 0.1 times 10,000 m, or 1000 m, equation (2-1) gives the worst case (i.e., conservative) predicted concentrations, C, derived as follows:

TABLE 2-1
Comparison of Characteristics of Dispersion Models for Routine
versus Accidental Releases: Standard EPA Models Such as ISC
Can Be Used for Routine Releases, whereas Vapor Cloud Dispersion
Models Must Be Used for Accidental Releases

ISC Model Characteristics as an Example of Routine Releases	Vapor Cloud Dispersion Models such as HGSYSTEM
Source well-defined (physical and chemical parameters easily measured).	Source poorly defined (physical and chemical parameters not all measured).
All gas or small particle release.	Release can be gas, liquid, or solid with wide range of droplet/particle sizes.
Continuous over time periods of about one hour or more.	Usually highly transient, with significant variations over periods of a few minutes.
Little thermodynamic interaction with the underlying surface.	Often strong thermodynamic interaction with underlying surface.
Phase changes and chemical reactions are minor effects.	Phase changes and chemical reactions can strongly influence vapor cloud behavior.
Generally one hour averaging time.	Variety of averaging times, from one second to several minutes, depending on toxicological or flammable effects.
No dense gas effects as determined by Ri criterion (see Equations 5-1 through 5-4).	Possible dense gas slumping.
Concentration criteria are generally small (a few ppm).	Concentration criteria can be large (several thousand ppm).

$$C = \frac{(0.1\ \text{kg}/600\ \text{s})}{(1\ \text{m/s})(50\ \text{m})(1000\ \text{m})} = 3.33 \times 10^{-9}\ \text{kg/m}^3 \qquad (2.2)$$

The concentration of 3.33×10^{-9} kg/m^3 is equivalent to about 3×10^{-3} ppmv, which is more than a factor of 10,000 below the concentration associated with health effects. In this case, the concentrations predicted by the worst-case screening approach are so far below the levels of interest that it can be concluded that more advanced modeling is unnecessary.

2.3. Scenarios Selected for Worked Examples

There are many vapor cloud dispersion models available, with a great variation in capabilities and accuracies. One of the objectives of the present book is to step through the calculations one might typically make for several specific release scenarios using publicly available formulas and computer models. Note that the use of any particular modeling method in this *Guidelines* book does not constitute endorsement by the AIChE/CCPS.

The CCPS Vapor Cloud Modeling subcommittee pooled their experience to generate the list of seven release scenarios, which cover the following source types:

- Dense aerosol jet from a pressurized Cl_2 liquid tank (Scenario 1)
- Evaporation from a cold boiling pool of Cl_2 (Scenario 2)
- Evaporation from a nonboiling pool of acetone (Scenario 3)
- High-speed vertical dense gas jet of a monocomponent, normal butane (Scenario 4)
- Venting of warm H_2SO_4 from building vent (Scenario 5)
- High-speed dense gas jet of multicomponents (Scenario 6)
- Short-duration time-variable HF release (Scenario 7)

For each model and each scenario, a base case and three sensitivity runs are carried out, with variations in input parameters such as roughness length, wind speed, and stability. The results of the analyses are given in several tables and figures, including, for example, predicted maximum cloud centerline concentrations as a function of downwind distance. Appendices A through G provide details on the applications of the models to the worked examples and the interpretation of the results. The general philosophy is to cover as broad a range of typical source types as possible and to include several examples for each of the commonly available dense gas models (ALOHA, DEGADIS, HGSYSTEM and SLAB). In addition, two of the EPA models for positively or neutrally buoyant gases (ISC and INPUFF) are included.

3

Input Data Required

Source emissions, meteorological conditions, site characteristics, and receptor requirements must be well-known in order to assure optimum performance by any model. The discussions below provide an overview of the types of input data required by vapor cloud models. In the section on meteorological input data, a set of equations is suggested for calculating some input parameters that are not generally well understood. The input data for the seven worked examples are listed in a Scenarios Data Archive (SDA) included later as Table 9-2.

3.1. Source Data

The source emissions model (see Section 4) requires several pieces of input information in order to operate, the most important of which is the total mass of the release (if instantaneous) or the mass emission rate (if continuous over some time period). The required source data are listed below:

a. *Physical and Chemical Properties of Material.* It is necessary to know the chemical composition of the liquid and/or gas, permitting the other properties to be determined. These may include the molecular weight, density, molecular diffusivity, conductivity, and boiling point. Temperature-dependent properties, such as vapor pressure, heat capacities, heat of vaporization, and surface tension may also need to be determined. If there are several components in a mixture, the properties of each component must be known. There are several useful reference documents that provide summaries of properties of many chemicals (e.g., Perry et al. 1984, AIHA 1995, NFPA 1994, NOAA 1992, and Urben 1995).

b. *Geometry of Pipe or Tank Source.* The release rate is also related to the geometry of the source, which includes the dimensions of the pipe, tank, or stack, and the characteristics of the source orifice. These parameters permit estimation of the release amount and type (e.g. a burst tank results in an instantaneous release, while a small hole in a tank results in a continuous release). The position of the orifice relative to the level of the liquid can determine whether the release is gaseous or liquid or a combination of both. There are clearly several types of source geometries, ranging from small diameter pipes to large tanks. It is also important to know the location of the source above the ground or whether part of the source is buried. If a pipe is involved, the height and the direction of discharge should be determined.

c. *Characteristics of Underlying Ground Surface for Liquid Spills and Ground-Based Dense Vapor Clouds.* Many hazardous releases are cold or dense gases or liquids which flow on the ground and exchange heat and moisture with the ground. At most chemical plants, dikes are built around storage tanks in order to contain spilled liquids. The dimensions of the dike must be input to the model. If there is no dike, the fluid will flow in a uncontrolled manner and the total evaporative emission rate could be larger. Clearly it is important to know the thermal conductivity, density, and specific heat of the ground, the moisture content, the porosity of the soil and terrain contours.

d. *Knowledge of Plant Operating Procedures.* Most plants have special safety procedures that influence the source release rate. For example, an operator may shut down a chemical reaction, or a gas pipeline may have valves which automatically close in the event of a rupture. More recently, many plants have implemented mitigation devices such as water spray curtains or vapor barriers. Each of these systems may limit the amount of the release and influence its variability with time.

e. *Time Variation.* Maximum predicted concentrations for a certain averaging time, T_a, are likely to be proportional to the maximum source emission rate over that time period. Consequently, knowledge of the time variability of the source emission rate would be useful. In some cases, the emission rate may show step changes as valves are closed or mitigation procedures implemented. Also, it is important to recognize that the flow rate from most pressurized containers decreases exponentially with time. If the release rate is highly variable in time, then along-wind dispersion should be accounted for in the transport and dispersion model.

f. *Presence of Gas and Aerosols in a Jet.* The fractions of the source emissions that are gases or aerosols imbedded in the jet must be prescribed. Also, some of the unflashed liquid may not be aerosolized but may spill onto the ground (i.e., rain-out). Recent model improvements for aerosol jets are described by AIChE/CCPS (1995). As an alternative, it is some-times possible to prescribe input conditions (e.g., "pseudo"-molecular weights) that indirectly account for the presence of the aerosols.

3.2. Site Characteristics

For model applications it is clear that a map of the site showing the possible source locations and the positions of nearby buildings, roads, and land-use types must be included in a set of input data. Topography is also useful, although most of the vapor cloud models assume flat conditions. Building dimensions in relation to the source position must also be input if the vapor cloud model can treat plume downwash behind buildings (e.g., GASTAR, CHARM, or ISC).

Most models require information on site roughness, since the intensity of ambient turbulence is affected by the surface roughness. The surface roughness parameter, z_0 (units m), which is derived from observations of wind speed, is about 1 m for cities, forests, and industrial complexes, 0.1 m for residential areas and agricultural crops, 0.01 m for grass, and 0.001 m for water or pavement surfaces. The roughness length is approximately 10% of the obstacle height for typical surfaces mentioned above. The 10% rule-of-thumb breaks down for very large roughness obstacles such as skyscrapers and mountains, which modify the wind flow in the lowest several hundred meters of the atmosphere but cannot be considered as minor perturbations to the boundary layer. The surface roughness length, z_0, used in the models should be restricted to a maximum value of 2 m. If the roughness elements are higher than the diffusing cloud, as they may be for a very dense cloud in an urban area or in a forest, then caution should be used in applying most hazardous gas models. Another special case occurs when an industrial site contains numerous buildings, pipelines, tanks, and other obstructions to the flow. In some cases it is necessary to use wind tunnels or water channels to determine the transport and dispersion of a plume released within these obstructions.

3.3. Meteorological Data and Formulas for Calculating Input Parameters

Modeling can be performed using either real-time input data or using historical or hypothetical conditions. Vapor cloud models such as CHARM, SAFER, ALOHA, and CHEM-MIDAS can be set up on-site at chemical plants so they can be run at the time an accident occurs in order to guide emergency response efforts. These models and other vapor cloud models, such as SLAB or PHAST, can also be used for design or planning purposes. Most modeling exercises now use historical or hypothetical situations for planning purposes. In the latter case, it is not important to precisely calculate the actual cloud trajectory based on wind direction observations. However, it is important to determine "worst-case" meteorological conditions. In any application, vapor cloud models require input of the wind velocity at some height (say, 10 m) and an indication of the atmospheric stability. These observations are best made on-site, but can also be obtained from nearby National Weather Service (NWS) sites, as long as the NWS site is reasonably representative of the facility being modeled. From this information and a knowledge of the roughness of the underlying surface, the model can estimate the velocity and entrainment rate of the vapor cloud.

An estimate of atmospheric stability is needed because the rate of turbulent dispersion of a passive gas plume is strongly dependent on stability. During sunny afternoons with light wind, the atmosphere is "unstable" and plume dispersion is at its maximum. During clear nights with light winds, the atmosphere is "stable" and plume dispersion is at its minimum. During cloudy or high-wind periods, the atmosphere is "neutral" and the plume disperses at an intermediate rate. Methods of parametrizing stability are given at the end of this section.

If the hazardous gas is cold or contains aerosols, it is possible that entrained ambient water vapor will condense within the cloud. Observations of ambient relative humidity are thus required to calculate this effect. All vapor cloud models also require input of air temperature near the ground.

Because many vapor cloud models require the input of meteorological variables such as the friction velocity, u_*, and a stability parameter known as the Monin–Obukhov length, L, and these variables are unfamiliar to most persons outside the boundary layer meteorology community, a brief review of these concepts is provided below.

The friction velocity, u_*, with units in meters per second, is a measure of the frictional stress exerted by the ground surface on the atmospheric flow. It is proportional to turbulent fluctuations in wind speed and is equal to about 10% of the average wind speed at a height of 10 m. The percentage increases as the surface roughness increases or as the boundary layer becomes more unstable. The friction velocity is used to parametrize the entrainment rate in most vapor cloud models.

The Monin–Obukhov length, L, with units in meters, is a measure of the depth of the air layer adjacent to the ground that is dominated by mechanical turbulence. The vapor cloud models use L in order to estimate cloud speeds and entrainment rates. If $|L|$ represents the absolute magnitude of L, then at heights z above $|L|$, the air flow is dominated by buoyancy effects. The value of L is positive during stable conditions (night time), is negative during unstable conditions (day time), and approaches very large values during neutral conditions (high-wind or cloudy). It is defined by the formula:

$$L = \frac{u_*^3}{[0.4g(H_f/T)]} \qquad (3\text{-}1)$$

where g is the acceleration of gravity, T is the absolute temperature, and H_f is the surface heat flux (defined to be positive downward). The scaling parameters known as the friction velocity, u_*, and the downward directed heat flux, H_f, with units m°K/s, can be expressed in terms of turbulent fluxes:

$$u_* = (-\overline{w'u'})^{1/2} \qquad (3\text{-}2)$$

$$H_f = -\overline{w'T'} \qquad (3\text{-}3)$$

where w', u', and T' are fast-response turbulent fluctuations in vertical wind speed (units m/s), longitudinal wind speed (units m/s), and temperature (units °K), respectively, and the overbars represent averages over time periods of about 10 to 60 minutes. These types of fast response measurements are made by research-grade instruments such as sonic anemometers.

In order to aid model users in prescribing input values for Monin–Obukhov length, L, Table 3-1 is provided. Typical values of L are given as a function of six familiar Pasquill–Gifford stability classes (A, B, C, D, E, F). The so-called Pasquill–Gifford stability class is used in many routine dispersion models including the EPA models such as ISC3 (EPA, 1995). Section 5.3 will provide tables for use in estimating the stability class as a function of easily measured parameters such as wind speed and cloudiness.

Together with the surface roughness length, z_0 (units m), the variables u_* and L can be used to express the vertical profile of wind speed, u (units m/s):

$$u = \frac{u_*}{0.4}\left(\ln\frac{z}{z_0} + 4.5\frac{z}{L}\right) \qquad \text{(near neutral and stable)} \qquad (3\text{-}4)$$

where z is height in meters. Equation (3-4) is valid for near-neutral and stable conditions. A similar formula is used for unstable conditions, but the z/L term is much more complicated (Hanna et al., 1982, Chapter 1). Because of the logarithmic term in equation (3-4), the data should fall close to a

TABLE 3-1
Relation between Wind Speed, u, Monin-Obukhov Length, L,
and Other Stability Descriptions

Description	Time and Weather	Wind Speed u	Monin-Obukhov Length L	Pasquill-Gifford Stability Class
Very Stable	Clear night	<3 m/s	10 m	F
Stable	↓	2 to 4 m/s	50 m	E
Neutral	Cloudy or Windy	any	>l100 ml	D
Unstable	↓	2 to 6 m/s	−50 m	B or C
Very Unstable	Sunny	<3 m/s	−10 m	A

straight line (see Figure 3-1) when observations of wind speed are plotted against logarithm of height during neutral conditions (i.e., when L is very large and the second term in the equation drops to zero). That straight line has a *slope* inversely proportional to the friction velocity, u_*, and *intercepts* the $u = 0$ axis, at $z = z_0$ (i.e., at the roughness length).

During stable and unstable conditions, it is more difficult to derive u_* from wind observations, since equation (3-1) shows that L is a function of u_*. For stable conditions equation (3-4) becomes a quadratic equation in u_* (see Hanna and Chang, 1992). For unstable conditions, an iterative procedure must be followed to obtain consistent estimates of u_* and L. The vapor cloud models contain algorithms that automatically solve these formulas and produce estimates of u_* and L for use within the models.

Many EPA passive-gas models such as ISC3 do not use equation (3-4) but instead estimate wind speed using a simple power law relation ($u \propto z^p$) where the power p is given as a function of Pasquill–Gifford stability class.

3.4. Receptor-Related Data

The locations of important fixed receptors such as monitoring sites, population centers, ignition sources, sensitive crops, and property boundaries should be considered part of the input data. In some cases, interest is focused not on specific locations, but on maximum concentrations at certain downwind distances.

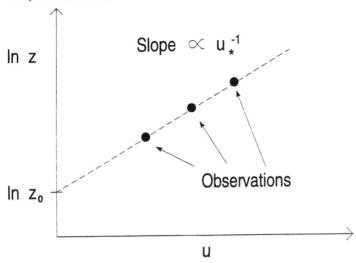

Figure 3.1. Schematic diagram of observed neutral wind profile, showing how surface roughness length, z_0, and friction velocity, u_*, can be estimated. During neutral conditions, $L \to \infty$, and equation (3-4) becomes $u = [(u_*/0.4) \ln (z/z_0)]$.

The concentrations of interest must be known for the particular chemicals being modeled. For example, these concentrations could be defined by toxicological information (as expressed by ERPG-1, 2, and 3) or by LFLs for flammable substances.

It is also necessary to specify the averaging time, T_a, which is defined as the length of time over which concentration data are averaged (see Section 6.1). This averaging time may be imposed by the characteristics of the monitoring instrument, by a toxicological criterion, or by a flammability criterion. For example, instantaneous averaging times are of interest for a flammable gas, while 10 minute averaging times may be appropriate for low concentrations of a toxic gas such as H_2S. Because of nonlinearities in relations between toxicological effects and averaging times, it may be necessary to consult a toxicologist if the averaging time is different from that listed in standard documents.

In some modeling studies, where interest is focused on certain concentrations defined by flammability or toxicity limits, contour maps are plotted containing those concentrations. The model could also produce integrated results such as total mass of flammable material.

Finally, the important question of indoor air pollution often arises. If the air exchange rate of the building can be determined, the indoor concentrations can be predicted based on the time series of the outdoor concentrations. A few preliminary models exist for this phenomenon, and one simple model is described and applied in Scenario 5 in Appendix E.

4

Source Emission Models

Modeling the source phenomenon for an accidental rupture or spill of hazardous material is perhaps the most critical step in the accurate estimation of downwind air concentrations resulting from the accidental release. The type of source emission calculations that are necessary for accidental releases are usually very different than the standard emission factors, such as in EPA's AP-42 document (EPA, 1995), that are used as input to routine dispersion or risk calculations. The emission calculations for accidental releases may involve multicomponents, two-phase (gas and liquid) releases, emission rates that vary dramatically in time, or the release of very dense, cold gases.

The units used to define the source emission term will depend on the scenario and the type of accidental release. Typically, emission rate units are in terms of mass per unit time (kg s^{-1}) for many types of sources. However, the emission term may have units of kilograms per square meter per second for evaporative emissions from a liquid spill; because these units must be multiplied by the pool area (m^2) for the total emission rate from liquid pools (in kilograms per second), calculation of pool spreading with time can be an important factor. The emission units may be in kilograms (i.e., total mass) for an instantaneous release; it is recognized that no release is truly "instantaneous," but for modeling purposes, it is convenient to represent releases with a duration of a few seconds as instantaneous.

The purpose of this chapter is to describe the basic physical and chemical principles that are appropriate in various types of rupture and spill emission situations, to enable an engineer to identify and understand the basic techniques used for calculating the source emissions from a particular type of accidental release. Wherever possible, analytical expressions or simplified empirical formulations for source terms are described, with emphasis on the most important parameters in the source term models. The emission cases in which more detailed computerized calculations are necessary will also be identified, for example, the classic chemical engineering problem of multicomponent phase equilibria.

This chapter is divided into three parts. Section 4.1 presents a summary of the conceptual process steps for determining the emission rate for an accidental release. The major part, Section 4.2, describes current analytical source emission models for the main types of accidental releases, including gas and liquid jet releases, two-phase flow, evaporation from pools, multicomponent evaporation, and cryogenic spill emissions. It should be stressed that many liquid spills are multicomponent and many emissions from tank or pipe ruptures are two-phase, and prediction schemes for these scenarios are not yet well-developed or tested. Finally, Section 4.3 describes the types of uncertainties that may affect the source term calculations.

Examples of applications of some of these source emissions models are given in the seven scenarios in Appendices A through G.

4.1. Conceptual Process for Source Term Determination

Accidental releases of hazardous materials can be of many different types—gas or liquid, instantaneous or continuous, from storage tanks or pipelines, refrigerated or pressurized, on land or water, confined or unconfined. In many cases, combinations of these scenarios may exist simultaneously. Storage tank or vessel accidental releases can result from corrosion, thermal fatigue, inlet or outlet pipe rupture, or valve failure.

Figure 4-1, reproduced from Fryer and Kaiser (1979), shows some of the different release mechanism scenarios that can occur with pressurized tanks and refrigerated liquids. Our current knowledge suggests that many emergency releases involve two-phase flows, and thus wherever vapor or liquid jets are indicated on Figure 4-1, the actual release could be a combination of vapor *and* liquid. In some of these cases, part of the liquid may "rain out" onto the underlying surface, and part may be entrained into the gas jet as an aerosol.

Obviously, there will be a large difference in the time variability of emissions and in the method of calculation for a catastrophic tank failure (instantaneous release) in comparison with a small puncture failure in a storage tank (continuous release). Also, different calculation techniques may apply depending on whether a tank failure occurs in the liquid region or in the vapor space above the liquid and whether the release contains one or two phases.

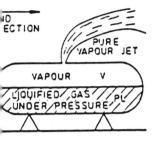

SMALL HOLE IN VAPOUR
SPACE-PRESSURIZED TANK

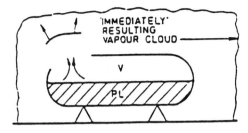

CATASTROPHIC FAILURE OF
PRESSURIZED TANK

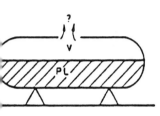

INTERMEDIATE HOLE IN
VAPOUR SPACE-PRESSURIZED
TANK

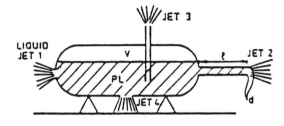

ESCAPE OF LIQUIFIED GAS FROM
A PRESSURIZED TANK

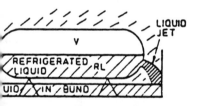

SPILLAGE OF REFRIGERATED
LIQUID INTO BUND

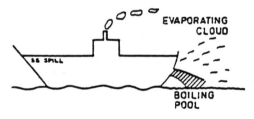

SPILLAGE OF REFRIGERATED
LIQUID ONTO WATER

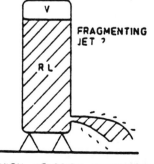

HIGH VELOCITY FRAGMENTING
JET FROM REFRIGERATED
CONTAINMENT

Figure 4-1. Illustration of some conceivable release mechanisms, from Fryer and Kaiser (1979). In most of these cases, the jet could be two phase (vapor plus entrained liquid aerosol). V = vapor, PL = pressurized liquid, RL = refrigerated liquids.

Figure 4-2 presents a flowchart of the conceptual process involved in determining a source emission rate for an accidental release. Four basic sequential steps are involved:

(1) Determine the time dependence of the emission release;

(2) Select the most applicable source term model for the situation;

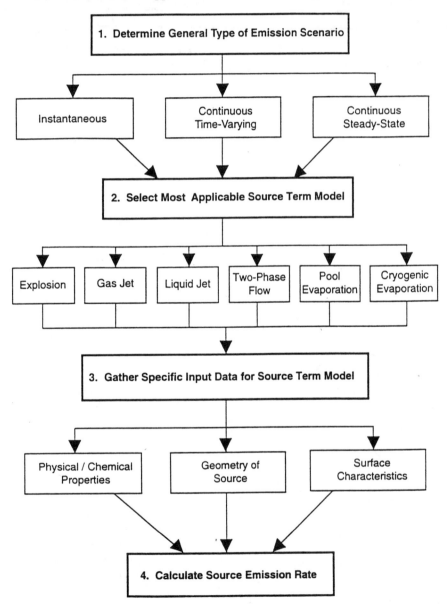

Figure 4-2. Conceptual process for source term determination.

(3) Gather specific input data and physical properties necessary for thesource term model; and

(4) Calculate the source emission rate.

As shown in Figure 4-2, accidental releases can be segregated into three general classes of time dependence: instantaneous releases; continuous but time-varying releases; and continuous steady-state or quasi-steady-state releases. This initial determination is important because different dispersion model techniques are used for these types of releases, and the calculated emission rate and units must be appropriate for the dispersion modeling technique. Thus, an initial knowledge of the general type of dispersion model that will be used is necessary when choosing a specific emission rate technique.

The choice of a specific emission model is dependent on the type of accidental release scenario and the physical properties of the chemicals being released. Tank or pipeline ruptures usually involve continuous time-varying releases that require gas jet, liquid jet, or two-phase jet emission models. In general, liquid pool evaporation also involves continuous time-varying releases because of spreading pool areas and time-dependent heat and mass balances on the liquid pool. For slow evaporation rates and a relatively constant pool area, a continuous steady-state liquid pool evaporation model may be appropriate. These emission calculation techniques are discussed in detail in Section 4.2.

Each type of source emission model requires different types of input information in order to calculate an emission rate. These source term model inputs were previously discussed in Section 3.1, and can be divided into three general classes: physical and chemical properties (e.g., vapor pressure); geometry of source (e.g., source orifice dimensions); and characteristics of the underlying ground surface (e.g., thermal conductivity of soil). An important surface characteristic is the presence of physical barriers (dikes) that can contain spilled liquids; if there is no dike, a liquid pool will spread over a much larger area, and total evaporative emissions will be larger.

4.2. Calculation of Source Terms

The purpose of this section is to describe the important physical processes in various types of releases, and to present the general analytical concepts that can be used to calculate the source emission rates from these releases. Where applicable, basic equations will be presented from standard references such as *Perry's Chemical Engineers' Handbook* (Perry et al., 1984, revised on a regular basis).

The following subsections review the physical processes relevant for the more important types of emission scenarios. Basic source emission models are described for gas and liquid jet releases, two-phase flow, evaporation from pools (cryogenic and nonboiling), and multicomponent evaporation. Some accidental release scenarios may require multiple emission models, or the emission models may need to be applied sequentially. For example, a two-phase jet from a tank rupture will release gas emissions to the air from the rupture point, as well as forming a liquid pool on the ground that will then evaporate.

The source models presented in the following subsections typically assume a simplified thermodynamic path (e.g., isentropic, isenthalpic) that governs the predicted temperature and pressure after release. An exact answer would require the solution of the combined steady-flow equations of mass, momentum, and energy. Thus, the assumption of a constant entropy or a constant enthalpy thermodynamic path is an approximation to the exact solution. At this time, little testing has been done to clarify our knowledge about the most appropriate thermodynamic path for predicting the initial cloud temperature for accidental releases to atmospheric conditions.

Pressurized single-phase gas jets are often treated in practice as isentropic expansions (i.e., reversible and adiabatic), which leads to predicted decreases in the initial cloud temperature and allows for significant increases in the kinetic energy (velocity) of the fluid. For gas releases with significant velocity changes on release, the isentropic approximation has been shown to be more accurate than an isenthalpic approximation (Britter, 1995). Section 4.2.1 describes the typical equations used for pressurized gas jet releases, assuming an ideal gas and an isentropic process.

For flashing jets that are primarily liquid at the exit plane, the large density of the fluid at the exit plane reduces the velocity change, and an isenthalpic expansion may be more appropriate. Through the use of the Joule–Thomson coefficient, an initial cloud temperature can be estimated. Here, a positive Joule–Thomson coefficient means the temperature drops during throttling, while an ideal gas would have a Joule–Thomson coefficient of zero. Unfortunately, few Joule–Thomson coefficients exist in the literature for practical application to accidental release analyses.

For situations such as flashing jets that are two-phase at the exit plane, the most accurate thermodynamic path is not clear. It may be reasonable in these cases to base the initial cloud temperature on using the results that produce the more conservative downwind impact for the problem being analyzed, unless sufficient information is available to justify the selection of one thermodynamic approach over the other in the first place. For example, both isenthalpic and isentropic thermodynamic paths might be used to estimate downwind cloud dispersion. Then, the initial cloud temperature that produces the more conservative result for the problem being analyzed should be used. Perhaps the most straightforward method to estimate the initial cloud temperature for both isenthalpic and isentropic expansions is from thermodynamic diagrams developed for common compounds that plot temperature, pressure, entropy, and enthalpy. Knowing the initial storage conditions of the compound as well as ambient pressure, the initial cloud temperature assuming isentropic or isenthalpic expansion can easily be determined from the diagrams. If such diagrams are not available, then more complex methods involving thermodynamic property estimation techniques or exact solutions to the equations of mass, momentum, and energy need to be employed.

4.2.1. Gas Jet Releases

One important type of accidental release is a gas jet from a small puncture in a pressurized gas pipeline or in the vapor space of a pressurized liquid storage tank, as shown in the top left drawing of Figure 4-1. Intuitively, one would expect to see an initial high release rate that decreases as the gas tank pressure decreases. When a small puncture occurs, gas can exit the puncture only as fast as its sonic velocity, because of a "choked" condition at the exit. Choked flow or critical flow simply means that the gas is moving through the puncture at its maximum possible speed, namely the speed of sound in the gas.

The emission rate equations for an ideal gas jet release are well known. The pressure criterion for choked or critical flow to occur, assuming an ideal gas exiting through a small hole, can be simply expressed (Perry et al., 1984):

$$\frac{p}{p_a} \geq \left(\frac{\gamma + 1}{2}\right)^{\frac{\gamma}{\gamma - 1}} \tag{4-1}$$

where

p = absolute tank pressure (N m^{-2})

p_a = absolute atmospheric pressure (N m^{-2})

γ = gas specific heat ratio (the heat capacity at constant pressure, c_p, divided by the heat capacity at constant volume, c_v)

For a typical gas, values of γ range from 1.25 to 1.5, and the critical p/p_a ratio, using equation (4-1), is approximately 2. Thus, if the tank pressure remains above approximately twice atmospheric pressure, the gas release rate will remain choked or critical.

If this choked flow condition is met, for an ideal gas exiting through an orifice under isentropic conditions, the gas emission rate will follow the critical flow relationship (Perry et al., 1984), which is independent of downstream pressure:

$$Q = c_0 A_h \left[p \, \rho_0 \, \gamma \left(\frac{2}{\gamma + 1} \right)^{\frac{\gamma + 1}{\gamma - 1}} \right]^{\frac{1}{2}}$$

$$(4\text{-}2)$$

where

Q = time-dependent gas mass emission rate (kg s^{-1})

c_0 = discharge coefficient for orifice (dimensionless)

A_h = puncture area (m^2)

ρ_0 = gas density in tank (kg m^{-3})

The coefficient of discharge (which typically ranges from 0.6 to slightly less than 1) accounts for reductions below the theoretical exit velocity due to viscosity and secondary flow effects. It depends upon nozzle shape and Reynolds number, and graphical relationships of c_0 for various types of orifices can be found in handbooks and fluid mechanics textbooks (e.g., Perry et al., 1984). For a conservative or maximum gas release rate, $c_0 = 1$ can be used in equation (4-2).

When the tank pressure decreases to the criterion in equation (4-1), or below about twice atmospheric pressure, choked flow no longer applies and the flow rate becomes subcritical (Perry et al., 1984):

$$Q = c_0 A_h \left(2 \rho_0 \, p \left(\frac{\gamma}{\gamma - 1} \right) \left[\left(\frac{p_a}{p} \right)^{\frac{2}{\gamma}} - \left(\frac{p_a}{p} \right)^{\frac{\gamma + 1}{\gamma}} \right] \right)^{\frac{1}{2}}$$

$$(4.3)$$

Note that equations (4-2) and (4-3) can be expressed in terms of tank temperature, T_0, through use of the formula $p = \rho_0 T_0 (R^*/M)$, where R^* is the universal gas constant and M is the molecular weight of the gas in the tank.

Although choked flow through a puncture remains sonic until the pressure drops to the criterion in equation (4-1), it is important to note from equation (4-2) that, even in a choked flow condition, the mass flow rate exiting the puncture will steadily decrease as pressure decreases. This is because the pressure (and thus the density) of the gas in the pipe or tank is decreasing with time. One will see initially very high mass flow rates that rapidly diminish with time in a roughly exponential decay manner.

To calculate the variations of pressure with time for use in equations (4-2) or (4-3), an assumption must be made whether the gas release occurs isothermally or adiabatically. A pipeline release can be approximately isothermal in cases with relatively small releases, because the expansion cooling is counteracted by frictional heating and heat transfer through the pipe walls. In other cases involving more rapid gas releases or insulated pipes, adiabatic conditions may be more appropriate. For the case of a gas jet exiting from a pressurized or cryogenic liquid storage tank, adiabatic conditions can be usually assumed for insulated storage tanks. The pressure then can be calculated from an iterative mass balance (i.e., the mass remaining in tank is equal to the initial mass minus the amount released), recognizing that the remaining gas pressure will depend on the adiabatic or isothermal assumption. For gas releases from pipe ruptures, the pipeline model of Wilson (1979, 1981) presents detailed equations, including pipe frictional losses, for determining the pressure and temperature to use in equation (4-2), for the assumption of isothermal or adiabatic flow. Wilson (1979, 1981) modified an empirical correlation developed by Bell (1978) to simplify the solution of isothermal, quasi-steady state flow from a gas pipeline. This solution applies to the mass of gas initially present in a pipe of length L_p and of cross-sectional area A_p. This empirical model expresses an isothermal pipeline gas release as a "double exponential" that decreases with time with two important time constants:

$$Q = \frac{Q_0}{(1 + \alpha)} \left(e^{-\frac{t}{\alpha^2 \beta}} + \alpha \, e^{-\frac{t}{\beta}} \right) \qquad (4\text{-}4)$$

where

Q = time-dependent gas mass emission rate (kg s^{-1})

Q_0 = initial gas mass emission rate at the time of the rupture (kg s^{-1})

α = nondimensional mass conservation factor

β = time constant for release rate (s)

t = time (s)

The mass conservation factor, α, in equation (4-4) is calculated as

$$\alpha = \frac{m_T}{\beta Q_o} \tag{4-5}$$

where m_T is the total mass in the pipeline (kg). The initial gas mass emission rate, Q_o, can be calculated by using the initial conditions in the pipe and either equation (4-2) or (4-3), for critical and noncritical flow, respectively.

The time constant, β, in equations (4-4) and (4-5) is a function of three nondimensional parameters:

$$K_F = \frac{D_p}{\gamma f L_p}; \quad K_A = \frac{A_h}{A_p}; \quad K_\gamma = \left(\frac{\gamma + 1}{2}\right)^{\frac{\gamma + 1}{\gamma - 1}} \tag{4-6}$$

where

$\begin{aligned}
D_p &= \text{pipe diameter (m)} \\
f &= \text{pipe friction factor (dimensionless)} \\
L_p &= \text{length of pipe (m)} \\
A_h &= \text{area of hole (m}^2) \\
A_p &= \text{cross-sectional area of pipe (m}^2) \\
\gamma &= \text{gas specific heat ratio (the heat capacity at constant} \\
&\quad\ \ \text{pressure, } c_p\text{, divided by the heat capacity at constant} \\
&\quad\ \ \text{volume, } c_v)
\end{aligned}$

The friction factor, which is a function of pipe roughness and the Reynolds number of the flow, can be derived from standard charts (e.g., Perry et al., 1984).

Using these parameters, Wilson developed a general equation for the time constant, β, which is valid for any hole size (NOAA, 1992):

$$\beta = \frac{2L_p K_F K_\gamma^{3/2}}{3cK_A^3}\left[\left(1 + \frac{K_A^2}{K_F K_\gamma}\right)^{3/2} - 1\right] \tag{4-7}$$

where

$$c = \text{speed of sound in the gas (m s}^{-1}) = \left(\frac{\gamma R^* T_o}{M}\right)^{1/2}$$

Equation (4-7) can be simplified for either very small or large holes in pipes. For small holes, where the factor $K_A^2/(K_F K_\gamma) \ll 1$, equation (4-7) reduces by a Taylor series approximation to

$$\beta \approx \frac{L_p\sqrt{K_\gamma}}{cK_A} \tag{4-8}$$

For large holes, where the factor $K_A^2/(K_F K_\gamma) > 30$, equation (4-7) reduces to

$$\beta \approx \frac{2L_p}{3c\sqrt{K_F}} \tag{4-9}$$

For a puncture in the vapor space of a pressurized liquid or cryogenic storage tank, the gas release is continually supplied by liquid evaporating (in the case of cryogenic liquids, boiling) in the tank, and the heat of vaporization of the liquid must be taken into account to calculate the temperature and pressure to use in equation (4-2). For an adiabatic system, the heat of vaporization is supplied primarily from the enthalpy of the liquid, and to a lesser extent from the enthalpy of the gas phase and the storage tank. Thus, in an insulated storage tank release, the liquid phase will become continually colder as more liquid evaporates and gas escapes. If the release takes place over a relatively long time period, heat transfer into the tank must also be considered. When all the liquid has evaporated, the release can be treated as a pure gas release.

4.2.2. Liquid Jet Releases

A puncture or failure in the liquid space of a pressurized or cryogenic storage tank can be considerably more complicated than the corresponding case of a vapor space failure. In the liquid release case, a liquid jet is typically propelled from the tank and, depending on the normal boiling point of the compound or the flashing behavior of a multicomponent mixture, some portion of the released liquid may instantaneously vaporize or "flash." In this subsection, the simplified and well-known case of a pure liquid release is examined; two-phase flow releases are described in Section 4.2.3.

Intuitively, one would expect that the release rate of a pure liquid jet would be proportional to the amount of pressure on the liquid in the tank, and would also depend on gravity because of the height of the liquid above the puncture. The equation typically used for calculating the liquid flow rate through a small puncture is based on the classical work of Bernoulli and Torricelli, and can be expressed as (Perry et al., 1984):

$$Q_l = c_o A_h \rho_l \left[2\left(\frac{p-p_a}{\rho_l}\right) + 2gH_l \right]^{1/2} \tag{4-10}$$

where

Q_l = liquid mass emission rate (kg s^{-1})

c_o = coefficient of discharge (dimensionless)

A_h = puncture area (m^2)

ρ_l = liquid density (kg m^{-3})

p = tank pressure (N m^{-2})

p_a = ambient pressure (N m^{-2})

g = acceleration due to gravity (9.81 m s^{-2})

H_l = height of liquid above the puncture (m)

The geometry of the tank or vessel can easily be incorporated into equation (4-10) by deriving geometry-specific values of H_l, the height of the liquid in the tank. For example, Wu and Schroy (1979) calculate the rate of liquid spill by using equation (4-10) and the following expression for the liquid height, H_l, for vertical cylindrical tanks:

$$H_l = \frac{4V_l}{\pi \, d_t^2} \qquad (4\text{-}11)$$

where

V_l = liquid volume remaining in the tank (m^3), and

d_t = tank diameter (m)

For horizontal cylindrical tanks, H_l and V_l are calculated by Wu and Schroy (1979) on a dynamic basis using the following equations:

$$H_l = \frac{d_t}{2}(1 - \cos \theta_l) \qquad (4\text{-}12)$$

$$V_l = \frac{L_t \, d_t^2}{4}\left[\theta - \frac{\sin(2\theta_l)}{2}\right] \qquad (4\text{-}13)$$

where

θ_l = the angle (in radians) relative to the upward vertical direction formed by the liquid surface remaining around the circumference of the tank

L_t = length of the horizontal tank (m)

Likewise, any other specific geometries (e.g., spherical) can be easily incorporated into equation (4-10).

Critical to the analysis of liquid pipeline failures are the valve spacing and closing time, which are coupled to determine the total volume of the spill. Long distances between the valves, or slow closing times, will allow a considerably greater volume of material to be released. A pipeline will empty its contents from the failure point until the first closed valve. Likewise, a storage tank failure will release liquid until the level falls below the puncture level; a gas phase release will then occur as described in Section 4.2.1.

4.2.3. Two-Phase Jet Releases

Recent research has shown that most emissions from pressurized liquid storage tanks are two-phase jets, consisting of a mixture of gas and liquid. Thus, the emission rate is somewhere between that for a gas and that for a liquid, and an estimate must be made of both the gas/liquid fraction and the combined gas and liquid flow rate in a flashing choked flow situation. The discussion that follows will consider relatively simple techniques for two-phase flow. For a more detailed discussion, AIChE's Design Institute for Emergency Relief Systems (DIERS) has published an excellent review of two-phase flow models (Fisher et al., 1992) and has developed a computer code, SAFIRE, that can consider multicomponent two-phase flow.

4.2.3.1. The Flashing Process
A fraction of a liquid spill may flash, depending on the normal boiling point of the liquid. Flashing is usually significant for any pressurized liquids, such as propane or butane, that have normal boiling points well below ambient temperature. One simple approach to determine the amount flashed is to assume that the vaporization process occurs so fast that it can be considered adiabatic. Thus, the heat of vaporization is provided solely from the enthalpy of the released liquid. For a single component and a constant temperature, this heat balance yields a simple expression for the fraction of liquid flashed:

$$\frac{Q_f}{Q_l} = \frac{c_p(T_o - T_b)}{H_{vap}} \tag{4-14}$$

where

Q_f = mass emission rate of liquid that flashes (kg s^{-1})

Q_l = total liquid mass emission rate (kg s^{-1})

c_p = heat capacity of the liquid (averaged between T and T_b) (J kg^{-1} K^{-1})

T_o = temperature of the liquid in the tank (K)

T_b = normal boiling point of liquid (K), assumed to be lower than T

H_{vap} = heat of vaporization of the liquid (J kg^{-1})

For binary or multicomponent mixtures, the flashing process is more complex and involves iterative flash calculations that require knowledge of the vapor–liquid equilibrium behavior of the mixture. The topic of multicomponent phase equilibria is a classic chemical engineering problem for distillation calculations, and standard numerical techniques have been developed for this procedure.

In the case of a pressurized liquid pipeline break, some of the fluid (liquid at the initial operating pressure) rapidly vaporizes or flashes to its gas phase as the pressure drops. As the pressure is relieved upstream, more liquid flashes to vapor. The process is very time-dependent; initially, at a short distance from the break, most of the upstream fluid does not realize that a break has occurred; Figure 4-3, reproduced from Drivas et al. (1983), shows this occurring pictorially. At longer times, the pressure wave has traversed farther upstream and more flashing occurs. Norris (1995) carried out experiments with several single and multiphase releases from pipelines and developed a model which can be used to determine whether the release will be all-liquid or flashing.

For both storage tank and pipeline liquid releases, as the liquid and flashed gases flow out of the puncture into the atmosphere, they will entrain ambient air, which both dilutes and warms the cloud. Because of high velocities and the vigorous flashing process, the liquid may exit to a large extent in droplet form. This process may result in considerably more material being transported downwind than would be calculated by adiabatic flash calculations. After the cloud becomes warmed by entrained air, more of the droplets will vaporize, again cooling the cloud.

A jet model that considers aerosol formation named RELEASE has been developed by the AIChE Center for Chemical Process Safety (CCPS), and an improved aerosol jet model with rainout, absorption, and reevaporation is currently under development with CCPS support (AIChE/CCPS 1995, Woodward 1995). The improved aerosol jet model accounts for droplet trajectories, plume trajectories in a cross wind, nonequilibrium drop temperature, liquid pool spreading, and evaporation from and dissolution to the pool. After correction of the experimental data for several variations in test conditions, the AIChE/CCPS (1995) model showed good agreement of predicted release rates with experimental releases of five compounds (water, Freon-11, cyclohexane, chlorine, and monomethyl amine).

4.2.3.2. Two-Phase Flow Emission Models
There are a number of empirical correlations that have been developed for two-phase flow situations. These methods are only preliminary, however, and current research has been focused on developing more comprehensive

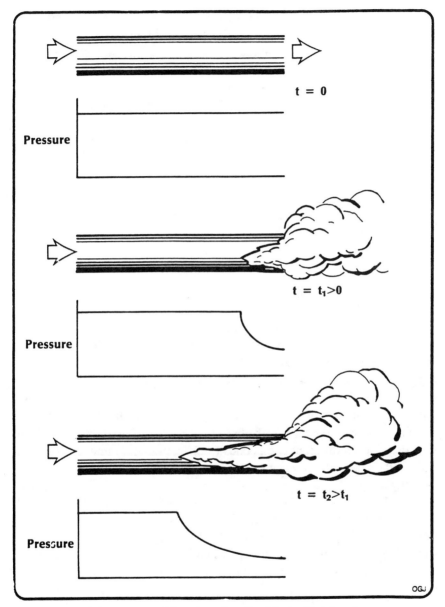

Figure 4-3. Pressure histories after liquid pipeline break.

models in the area of two-phase choked flow releases from pressurized tanks or pipes. Empirical models have been developed by several researchers for two-phase choked flow of initially saturated liquids, initially saturated vapors, and mixtures of liquids and vapors. Currently, two-phase flow empirical models can handle only single components; a multicomponent two-phase flow situation requires iterative numerical thermodynamic and heat balance calculations.

The most widely used technique for estimating two-phase flow for a one-component liquid–vapor mixture is the homogeneous equilibrium flow model (HEM). The HEM is based on four main assumptions:

- the liquid–vapor mixture is homogeneous;
- the liquid phase and the vapor phase are in thermal equilibrium; and
- the liquid phase and the vapor phase are moving at the same velocity;
- the expansion process is isentropic.

Using the HEM, the critical or choked flow rate for various initial conditions can be derived by using the basic energy balance equation for flow rate and determining where this flow rate reaches a maximum value (i.e., the choked flow rate) as a function of pressure.

Leung (1986), Leung and Grolmes (1988), and Sallet (1990) have used this HEM technique to develop two-phase source release models that involve relatively complex empirical parameters for one-component flashing choked flow. Sallet (1990) used normalized variables based on thermodynamic critical constants in his HEM analysis and claims an accuracy of his two-phase flow empirical correlations to within ±7%, based on comparison with data for ten different compounds (propane, water, ammonia, chlorine, carbon dioxide, difluoroethane, ethylene, Freon 22, sulfur dioxide, and trimethylamine).

Fauske and Epstein (1988) have derived relatively simple analytical expressions for two-phase choked flow. For subcooled stagnation conditions, where the stagnation pressure (the pressure at zero velocity) is substantially larger than the vapor pressure corresponding to the stagnation temperature, an all-liquid Bernoulli-type equation is suggested, similar to equation (4-10):

$$G \approx c_0 (2[p_0 - p_v(T_0)]\rho_l)^{1/2} \qquad (4\text{-}15)$$

where

G = mass flux (kg s^{-1} m^{-2})

c_0 = discharge coefficient (dimensionless)

p_0 = stagnation pressure (N m^{-2})

$p_v(T_o)$ = vapor pressure at stagnation temperature T_o (N m^{-2})

ρ_1 = liquid density (kg m^{-3})

For saturated liquid conditions, where $p_o = p_v(T_o)$, Fauske and Epstein (1988) suggest:

$$G \approx \frac{FH_{vap}}{v_{lg}\sqrt{Tc_{pl}}}$$

(4-16)

where

F = frictional loss factor

H_{vap} = heat of vaporization (J kg^{-1})

v_{lg} = difference in specific volume between liquid and gas (= $v_g - v_l$) (m^3 kg^{-1})

T = reservoir temperature (K)

c_{pl} = liquid heat capacity (J kg^{-1} K^{-1})

The frictional loss factor, F, accounts for frictional dissipation based on the length-to-diameter (L/D) ratio of the exit tube. Fauske and Epstein (1988) suggest $F \approx 1$ for $L/D = 0$, $F \approx 0.85$ for $L/D = 50$, $F \approx 0.75$ for $L/D = 100$, $F \approx 0.65$ for $L/D = 200$, and $F \approx 0.55$ for $L/D = 400$.

Finally, for the transition region from subcooled to saturated stagnation conditions, Fauske and Epstein (1988) suggest a combination of equations (4-15) and (4-160):

$$G \approx c_o \left(2[p_o - p_v(T_o)]\rho_1 + \frac{F^2 H_{vap}^2}{v_{lg}^2 T c_p} \right)^{1/2}$$

(4-17)

Researchers find that a pipe length on the order of 10 cm or greater is required between the source vessel and the emission point for the two-phase flow to be developed to an equilibrium stage (Fauske, 1985). It is stressed that these two-phase flow equations have not yet been tested against a wide range of chemicals and conditions. More importantly, there are very few correlations currently available to estimate air entrainment into a two-phase momentum jet. In addition, it should be pointed out that equations (4-15) through (4-17) were originally derived in order to conservatively design relief valves, and that the predicted mass fluxes may not be conservative from the point of view of estimating downwind concentrations.

Alternate approaches to the two-phase flow problem are presented by, for example, Webber and Kukkonen (1990) and Woodward (1993), who discuss detailed analyses of the problem of expansion zone modeling involving a high momentum, two-phase, choked flow situation.

4.2.4. Liquid Pool Spreading

The cases of gas, liquid, and two-phase jet releases that have been described above can be classified as continuous time-varying releases, where the release rate is typically decreasing with time. These types of failures result from punctures in storage tanks or pipelines. Also of importance is the case of a failure of a vessel or tank that results in the formation of a liquid pool. Because of the obvious combustion danger of large liquefied natural gas (LNG) spills, a considerable amount of effort has been spent on modeling the spreading and evaporation of cryogenic liquid pools. The subject of gaseous emissions due to evaporation of nonboiling liquid pools is also important, where the evaporative emission rate may be lower than for cryogenic spills but the toxicity of the emitted compounds may be significant.

Liquid spills on water, as well as unconfined spills on land, will generally increase in area with time; this increase in pool area is caused primarily by gravity spreading. Because the total evaporative emission rate from liquid pools (in kg s^{-1}) increases approximately linearly with the area of the liquid surface, calculation of pool spreading with time can be an important factor in estimating the total amount of evaporative emissions released to the atmosphere.

Shaw and Briscoe (1978) present a good summary of the equations used to calculate how unconfined spills increase in area on both land and water. Their model was derived for cryogenic spills and assumes that, for spills on land, heat transfer into the boiling liquid pool is limited by conduction from the soil. A set of four simultaneous heat and mass conservation equations were solved analytically by Shaw and Briscoe (1978) for the gravitational phase of spreading to determine the pool area increase with time. Pool spreading results were derived for instantaneous and continuous spills on land and also for instantaneous and continuous spills on water. The Shaw and Briscoe (1978) pool spreading equations can be summarized as:

Case 1. *Instantaneous spill on land*

$$r = \left[\left(\frac{8gV_0}{\pi} \right)^{1/2} t + r_0^2 \right]^{1/2} \tag{4-18}$$

Case 2. *Continuous spill on land*

$$r = \sqrt{\frac{2}{3}} \left(\frac{8gB_1}{\pi} \right)^{1/4} t^{3/4} \tag{4-19}$$

Case 3. *Instantaneous spill on water*

$$r = \left[\left(\frac{8g(\rho_w - \rho_l)V_o}{\pi \rho_w} \right)^{1/2} t + r_o^2 \right]^{1/2} \qquad (4\text{-}20)$$

Case 4. *Continuous spill on water*

$$r = \sqrt{\frac{2}{3}} \left(\frac{8g(\rho_w - \rho_l)B_l}{\pi \rho_w} \right)^{1/4} t^{3/4} \qquad (4\text{-}21)$$

where

r = pool radius (m)

g = acceleration due to gravity (9.81 m s^{-2})

V_o = volume of instantaneous spill (m^3)

t = time after initial spill (s)

r_o = initial radius of contained liquid (m)

B_l = continuous liquid spill rate (m^3 s^{-1})

ρ_w = density of water (kg m^{-3})

ρ_l = density of spilled liquid (kg m^{-3}), which must be less than the density of water

The above equations (4-18) through (4-21) from Shaw and Briscoe (1978) ignore any viscosity and surface tension effects and assume that the spread of the pool is due entirely to the conversion of gravitational potential energy into kinetic energy. Thus, these equations may not be applicable for very thin liquid layers or very viscous liquids (see Wu and Schroy, 1979, for a discussion of how to model the spread of viscous liquids). Also, these equations are for spills on flat terrain. Pool spreading is likely to be larger for sloping terrain because the initial pool will flow downhill to cover a broader area.

Generally, storage tanks on land normally have some form of confinement around them in order to limit the spread of the liquid in the event of an accident. This may be a dike, a dirt embankment around the tank, or in some cases, the area surrounding the tank is dug out to hold the tank contents in the event of a failure. This confinement can be extremely important in limiting the emission rate from a storage tank failure, since the total evaporative emission rate is approximately proportional to the area of the spill. Figure 4-4 shows that limiting the pool area is a very effective method to limit the source emission rate. These curves, from Shaw and Briscoe

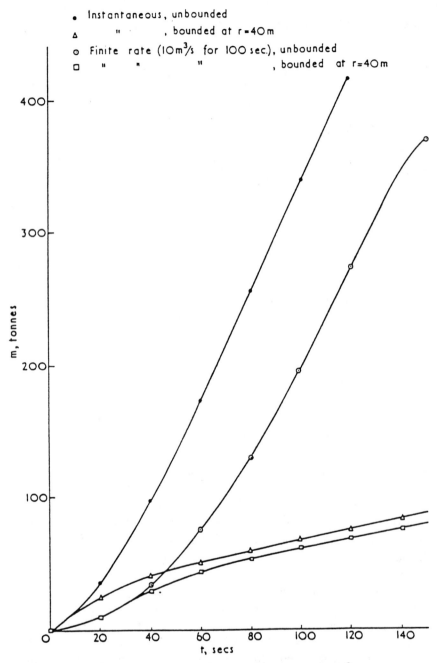

Figure 4-4. Calculation of mass vaporized M at time t for a spill of 1000 m^3 (416 tons) LNG on average soil, from Shaw and Briscoe (1978). A large difference between bounded and unbounded spills is evident.

(1978), show how source emissions vary as a function of time for both confined and unconfined cases of a 1000 m³ LNG spill on land. For this particular example, the unconfined source emissions at a given time are roughly a factor of five times higher than the confined emissions.

4.2.5. Liquid Pool Evaporation

When a major cryogenic accident occurs, where the normal boiling point of a liquid such as chlorine is below ambient temperature, the fluid will pour out quickly on the ground forming a large and cold liquid pool. The emission release in this scenario will result in a cold, dense gas cloud that hovers over the boiling liquid pool.

The evaporation rate from cryogenic pools depends almost entirely on the heat balance affecting the boiling liquid pool. It can be assumed that any heat transfer into the liquid pool will be used to evaporate more gas, at a rate depending on the input heat transfer rate and the heat of vaporization of the liquid.

For nonboiling liquid pools such as acetone or crude oil spills, the evaporation rate may be controlled by meteorological variables such as wind speed (gas-phase resistance) or by diffusion through the liquid layer (liquid-phase resistance). In either case, the liquid temperature is an important variable in calculating the evaporative emission rate of nonboiling liquid pools. Thus, the heat balance of a liquid pool is important not only for the evaporation of cryogenic spills, but for nonboiling liquid pools as well.

Figure 4-5, reproduced from Shaw and Briscoe (1978), shows the general heat balance of a liquid spill. Typically, the primary source of heat

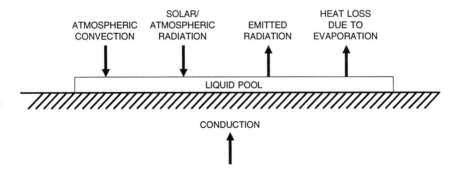

Figure 4-5. The heat budget of an evaporating pool (Shaw and Briscoe, 1978).

in the initial stage of evaporation is conduction through the ground or soil. Convection from the air, incident solar energy, evaporative heat loss, and radiative heat transfer also contribute to the heat balance of an evaporating pool. The rate of change of the temperature of a liquid pool can be expressed as the combined effect of individual heat transfer components,

$$mc_p \frac{dT_p}{dt} = q_{cond} + q_{conv} + q_{sun} + q_{rad} - q_{evap} \qquad (4\text{-}22)$$

where

T_p = pool temperature (K) at any given time t

m = liquid mass in pool (kg)

c_p = heat capacity of liquid (J kg^{-1} K^{-1})

q_{cond} = conduction heat transfer from ground (J s^{-1})

q_{conv} = convection heat transfer from air (J s^{-1})

q_{sun} = incident solar radiation (J s^{-1})

q_{rad} = radiative heat transfer from air (J s^{-1})

q_{evap} = heat loss due to evaporation (J s^{-1})

Because the liquid mass in the pool changes with time, the corresponding mass balance equations can be solved to accurately determine the temperature behavior and evaporation rate of a liquid pool as a function of time (e.g., Brighton, 1990; Cavanaugh et al., 1994; Leonelli et al., 1994). The specific heat transfer formulations for conduction, convection, solar radiation, and radiation are usually based on standard heat transfer relations.

Conduction heat transfer from the ground or soil, which is the primary source of heat for cryogenic spills, is discussed in detail in the next subsection. To calculate convective heat transfer from the air in equation (4-22), the standard approach is

$$q_{conv} = c_{HT} A_p (T_{air} - T_p) \qquad (4\text{-}23)$$

where

c_{HT} = heat transfer coefficient (J s^{-1} m^{-2} K^{-1})

A_p = pool area (m^2)

T_{air} = ambient air temperature (K)

The convective heat transfer coefficient, c_{HT}, can be derived from chemical engineering correlations for heat and mass transfer over flat plates (e.g., Perry et al., 1984).

The radiative heat transfer in equation (4-22) is typically modeled using the standard relationship for long-wave atmospheric radiation,

$$q_{rad} = \sigma_{SB} \varepsilon A f_{area}(T_{air}^4 - T_p^4) \tag{4-24}$$

where

σ_{SB} = Stefan-Boltzmann constant (5.67×10^{-8} J s^{-1} m^{-2} K^{-4})

ε = liquid emissivity (dimensionless)

f_{area} = area view factor (dimensionless)

The area view factor is dependent only on geometrical considerations, and standard charts of this factor can be found in handbooks and heat transfer textbooks (e.g., Perry et al., 1984).

The incident solar radiation, q_{sun}, can be modeled using the approximation derived by Raphael (1962):

$$q_{sun} = 1111 A_p(1 - 0.0071 C_I^2)(\sin \varphi_s - 0.1) \tag{4-25}$$

where

C_I = dimensionless cloudiness index (0 for clear day, 10 for complete cover)

φ_s = solar altitude (degrees above the horizon)

The "solar constant" is 1111 J s^{-1}m^{-2} and the pool area, A_p, must have units m^2. Equation (4-25) is valid only for sin φ_s greater than 0.1 (i.e., at low sun angles where sin $\varphi_s < 0.1$, the incident solar radiation can be assumed to be 0.0). The solar altitude can be derived as a function of latitude, longitude, hour of the day, and Julian day of the year (Raphael, 1962). For example, the EPA's ISC model (EPA, 1995) automatically calculates φ_s.

The heat loss by evaporation is simply

$$q_{evap} = Q_a H_{vap} \tag{4-26}$$

where

Q_a = time-dependent evaporation rate to the air (kg s^{-1})

H_{vap} = heat of vaporization of the liquid (J kg^{-1})

It should be noted that most spill models assume that the temperature in the liquid pool is uniform vertically, which may not be a good assumption for relatively deep liquid layers associated with very large liquid spills. Studer et al. (1988) have developed a two-compartment liquid layer model for cryogenic spills, consisting of a thin surface layer and a bulk liquid pool layer. They find that the surface layer can be subcooled, relative to the bulk liquid, in the initial time period of release, and that the evaporation rate is thus decreased because of the significant cooling of the surface layer.

4.2.5.1. Limiting Case: Cryogenic Pool Evaporation

For cryogenic spills, the primary source of heat input for evaporation is conduction through the ground or soil. Thus, equation (4-22) can be simplified by considering only the conduction heat transfer term (q_{cond}) and the evaporative heat loss (q_{evap}). The pool temperature for cryogenic spills will rapidly decrease and remain near its boiling point, and any further evaporation will depend primarily on the heat input from conduction,

$$q_{evap} = Q_a H_{vap} \approx q_{cond} \qquad (4\text{-}27)$$

The conduction heat transfer term from the ground is usually calculated by solving the standard one-dimensional heat flow equation with appropriate boundary conditions. Because the heat of vaporization is provided primarily by conduction heat transfer, the ground below a boiling cryogenic liquid normally becomes colder, and ice formation in wet soil is a possibility. The cold ground temperatures, and especially ice formation in wet soil, can significantly decrease heat transfer to the pool, and thus decrease the evaporation rate. Generally, infiltration into the soil surface is neglected for cryogenic liquids, because of the frost layer rapidly established beneath the surface.

To calculate the conduction heat flow rate into a boiling pool from the ground, most models use a standard analytical solution of the one-dimensional heat transfer equation (Carslaw and Jaeger, 1959):

$$q_{cond} = \frac{2k_s A_p (T_s - T_b)}{\sqrt{\pi \alpha_s t}} \qquad (4\text{-}28)$$

where

q_{cond} = conduction heat flow into the pool (J s^{-1})

k_s = thermal conductivity of the soil (J s^{-1} m^{-1} K^{-1})

A_p = pool area (m^2)

T_s = soil temperature (K)

T_b = boiling point of the liquid (K)

α_s = thermal diffusivity of the soil (m^2 s^{-1})

t = time (s)

Corrections can be made, if necessary, to k_s and α_s in this expression to account for the presence of water or ice in the soil. As an example of the dependence of the liquid boiling point on the evaporation rate, Figure

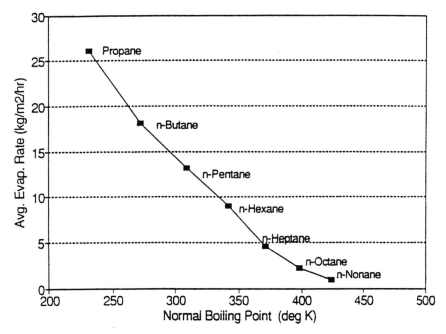

Figure 4-6. Theoretically predicted average evaporation flux as a function of chemical normal boiling point (Cavanaugh et al., 1994).

4-6, reproduced from Cavanaugh et al. (1994), shows the calculated cryogenic spill evaporation rate as a function of the boiling point for seven different compounds.

4.2.5.2. Limiting Case: Nonboiling Liquid Pool Evaporation

In the case of liquid spills where the normal boiling point of the liquid is above ambient (i.e., noncryogenic), a liquid pool will form on the ground and will slowly evaporate at a rate depending on its vapor pressure, pool area, and atmospheric conditions. Most hydrocarbon spills (for example, acetone) would fall into this category.

For the case of a slowly evaporating liquid pool, most emission models assume that the evaporation is limited by mass transfer from the liquid pool to the gas phase. The evaporation rate for a single component liquid, assuming an ideal gas phase, a well-mixed liquid phase (i.e., either a very shallow liquid layer or a well-mixed deeper pool), and a steady-state condition, can be expressed as (Fleischer, 1980):

$$Q_a = \frac{k_g A_p p_s M}{R^* T_i}$$

$$(4\text{-}29)$$

where

Q_a= evaporative emission rate to the air (kg s^{-1})

k_g = mass transfer coefficient (m s^{-1})

A_p= pool area (m^2)

p_s = vapor pressure of the compound at temperature T (N m^{-2})

M = molecular weight (kg mole^{-1})

$R*$= gas constant (8.31 joules mole^{-1} K^{-1})

T_i = liquid–gas interface temperature (K)

This equation is conservative for rapidly evaporating liquids, since it assumes that no material is present in the bulk air above the liquid. For simplification, the temperature in equation (4-29) can be considered equal to ambient (i.e., evaporation is so slow that the liquid remains at ambient temperature); however, for better accuracy, the liquid temperature should be calculated from the complete heat balance in equation (4-22).

To calculate the mass transfer coefficient, k_g, in equation (4-29), there are two main methods that are used in source emission models. One is a chemical engineering approach, which uses the general relationship (Perry et al., 1984):

$$N_{Sh} = \frac{k_g d_p}{D_m} = f(N_{Sc}, N_{Re})$$

(4-30)

where

N_{Sh} = Sherwood number

d_p = effective pool diameter (m)

D_m = molecular diffusivity of the compound (m^2 s^{-1})

N_{Sc} = Schmidt number (= kinematic viscosity divided by the molecular diffusivity, D_m)

N_{Re} = Reynolds number

In this approach, standard chemical engineering empirical correlations, based on laminar or turbulent flow over flat plates, are used to calculate k_g in equations (4-29) and (4-30). For example, Fleischer (1980) uses a standard flat plate correlation for the mass transfer coefficient, assuming that the laminar to turbulent flow transition occurs at a Reynolds number (based on pool diameter) of 320,000:

$$N_{Sh} = 0.664 N_{Sc}^{1/3} N_{Re}^{1/2} \qquad\qquad N_{Re} < 320,000 \qquad (4\text{-}31)$$

$$N_{Sh} = 0.037 N_{Sc}^{1/3} [N_{Re}^{0.8} - 15200] \quad N_{Re} > 320,000 \qquad (4\text{-}32)$$

The other main method to calculate the mass transfer coefficient, k_g in equation (4-29), for slowly evaporating pools is a meteorological approach based primarily on the work of Sutton (1953), who solved the steady-state atmospheric diffusion equation over a liquid pool with a power-law velocity profile and a corresponding power-law eddy diffusivity profile. An expression for k_g as a function of windspeed and liquid pool dimension has been developed by Mackay and Matsugu (1973), who formed a correlation based on experimental data on evaporation for neutral atmospheric stability:

$$k_g = 0.00482 N_{Sc}^{-0.67} u^{0.78} d_p^{-0.11}$$

                                                                              (4-33)

where

k_g = mass-transfer coefficient (m s^{-1})

N_{Sc} = Schmidt number

u = wind speed (m s^{-1})

d_p = diameter of pool (m)

This expression for the mass transfer coefficient in equation (4-33) has been commonly used in many source emission models to calculate k_g for the evaporation rate from single-component, well-mixed liquid pools (e.g., Cavanaugh et al., 1994). Kawamura and Mackay (1987) present some more recent experimental data that also validates this expression; they found that the differences between the predictions of equation (4-33) and experimental evaporation data of seven different compounds ranged from –35% to +47%, relative to the experimental value. As an example of the use of equations (4-29) and (4-33) to calculate evaporation rates, Figure 4-7, reproduced from Cavanaugh et al. (1994), shows the parametric effect on the evaporation rate as a function of ambient temperature and wind speed; in these

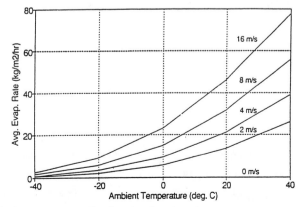

Figure 4-7. Average evaporation flux as a function of ambient temperature and wind speed (Cavanaugh et al., 1994).

calculations, the complete heat balance in equation (4-22) was used to calculate the pool temperature.

However, in the initial stages of a large spill or for relatively deep liquid pools, diffusion through the liquid (liquid-phase resistance) may be the rate-limiting step for components that have a relatively high vapor pressure. For very volatile compounds and a relatively thick or viscous liquid layer, the evaporation rate for these compounds can be limited by how fast they can diffuse to the surface of the liquid layer.

Several researchers have considered the problem of liquid-phase resistance in liquid pool evaporation. Wu and Schroy (1979) present a pool evaporation model that incorporates liquid-phase resistance by the use of a liquid-phase mass transfer coefficient. Kawamura and Mackay (1987) have developed an empirical liquid-phase "resistance factor" that increases with the compound's boiling point. Hanna and Drivas (1993) present a method for determining the rate-limiting step, whether liquid diffusion or mass transfer to the air, for calculation of liquid spill evaporation. In the application of their model to oil spill evaporation, Hanna and Drivas (1993) found that diffusion through a thin but viscous oil layer was the controlling factor for highly volatile compounds such as benzene and toluene, but that gas-phase mass transfer was the controlling factor for less volatile compounds (C_8 and higher carbon-number hydrocarbons).

4.2.6. Multicomponent Evaporation

Many potentially hazardous liquids are multicomponent, and hence, cannot be treated by single component equations. Drivas (1982) extended the equations for single components to derive equations for multi-components. Assuming that q_a is the total evaporative emission flux (kg m^{-2} s^{-1}) of all components, the following empirical equation can be derived:

$$q_a = \frac{k_g \, m_T^0}{n_T \, RT} \frac{\displaystyle\sum_{i=1}^{N} x_i^0 \, p_{is} \, M_i \, e^{-k p_{is} t}}{\displaystyle\sum_{i=1}^{N} x_i^0 \, M_i} \qquad (4\text{-}34)$$

where

k_g = mass transfer coefficient for the gas phase defined by equation (4-30) or (4-33),

m_T^0 = total initial mass of evaporable liquid (kg)

n_T = total moles of liquid

x_i^0 = initial liquid mole fraction of component i

p_{is} = saturation vapor pressure of component i (N m^{-2})

M_i = molecular weight of component i (kg mole^{-1})

k = $k_g A_p/(n_T RT)$

R = universal gas constant (8.31 J mole^{-1} K^{-1})

The initial fraction x_i^0 can be derived from a knowledge of the initial liquid composition or from gas-phase observations over the liquid, using Raoult's or Henry's law. It is assumed that mass transfer in the liquid is sufficiently rapid that total transfer is dominated by the gas phase. Drivas (1982) has verified this equation by comparisons with observations of evaporation of C_8 to C_9 straight-chain hydrocarbons from a multicomponent oil spill. However, there is clearly much more work required to refine these procedures and incorporate them into applied models.

A few source models consider multicomponents numerically by keeping account of the mass balances of each component over small time steps; recent examples are the Cavanaugh et al. (1994) and the Leonelli et al. (1994) multicomponent liquid spill models. A simpler model that can handle binary liquid pool evaporation is that of Wu and Schroy (1979). Their model is equivalent to equation (4-34) for N equal to 2, and they also account for resistance to mass transfer in the liquid phase by use of a liquid-phase mass transfer coefficient. Wu and Schroy (1979) have validated their model using literature data for mixture evaporation.

4.3. Uncertainties in Source Term Estimation

Source emissions, as well as meteorological conditions, can be highly variable and must be well-known in order to assure optimum performance by any model. If the concentration is directly proportional to a source input parameter, then any uncertainty in that parameter directly propagates to the same uncertainty in the predicted concentration. Thus, any inaccuracies in the source emission estimation will greatly influence the subsequent dispersion calculation of air concentrations; typically, the error in concentration prediction is directly proportional to the error in mass release rate.

There are three basic causes of inaccuracies in source term estimation:

- errors in defining the type of source release;
- inaccuracies in estimating the input parameters needed for a specific source term model; and
- the basic uncertainty of the source emission model itself.

All three of these types of inaccuracies will contribute to the overall accuracy of the source term estimation and the subsequent dispersion calculations.

For example, if the source were treated as an instantaneous point source rupture, the predicted concentrations would be quite different than if the source were treated as a continuous area source. Also, because of the nature of accidental spills, it is possible that some key source term parameters, for example, the volume of spilled material, can be only roughly estimated.

Clearly, some source term models are more reliable than others. For example, single-component pure liquid or gas jet equations have been well-known for many years, while two-phase flow prediction schemes are still in the research stage. A number of calculational techniques for estimating source emission rates from various types of accidental releases have been summarized in Section 4.2 above. If ranked by degree of accuracy, ranging from the most well-known emission techniques to the most uncertain, a list might be generated in the following order:

- liquid jets
- gas jets
- liquid pool evaporation—single component
- cryogenic liquid spills
- liquid pool evaporation—multicomponent
- two-phase releases—single component
- two-phase releases—multicomponent

The most well-known emission technique is the simple pressure-dependent and gravity-dependent pure liquid jet equation, which is based on the classic work of Bernoulli and Torricelli. On the side of highest uncertainty, the emissions from many liquid spills are multicomponent, and many accidental releases from tank or pipe ruptures are two-phase; prediction schemes for these scenarios are not well-developed or tested. The formulation of calculational techniques that can estimate the fraction of liquid, gas, and aerosol resulting from an accidental release is an active area of current research.

5

Dispersion Models

Chapter 4 covered the source emissions aspects of the generalized hazardous gas modeling diagram that was previously drawn in Figure 2-1. This section will cover all aspects of transport and dispersion of the released material once it is in the atmosphere. It is assumed that all localized effects (e.g., thermodynamics effects associated with jet expansion to atmospheric pressure) near the source opening are important only in those first few meters near the source, and have negligible effect on ground level concentrations at downwind distances, x, of 100 m or more. Generally, the primary interest will be in calculating one or more of the following quantities:

- Distances to certain toxicological concentration levels at ground level
- Contour plots of these concentrations
- Flammable mass within certain concentration bounds.

Note that toxicological concentration levels are always defined in terms of an averaging time, T_a. In some applications, dosages, or time-integrated concentrations, are also of interest.

5.1. Critical Richardson Number Criterion

Emphasis in Chapter 5 will be on dense gas releases, although formulas for passive gas releases will be briefly reviewed. Whether or not a cloud should be considered dense depends on its excess density, its volume or volume flux, and ambient conditions, as expressed by the Richardson number defined below. The excess cloud density affects the cloud in two primary ways—(1) the entrainment rate of ambient air into the top of the cloud is reduced, and (2) the rate of lateral cloud spreading is enhanced (Britter,

49

1989). It is appropriate to apply dense gas models if the initial potential energy of the dense gas cloud is significant when compared with the ambient atmospheric kinetic energy. Otherwise, standard EPA models such as ISC3 (EPA, 1995) should be applied to neutrally or positively buoyant clouds. When analyzing ground-level releases of dense gases, the Richardson number criterion is often used, where the initial Richardson number, Ri_0, represents the ratio of the potential energy due to density excesses inside the dense cloud to the kinetic energy due to ambient turbulence. If the value of Ri_0 is smaller than some critical number, then the cloud's motion is dominated by ambient turbulence and the dense gas effects are relatively unimportant. Analyses of vertical entrainment rates into dense gas clouds at ground level in laboratory experiments suggest the critical value of Ri_0, as defined below, is about 50 (see plots of entrainment rates published by Spicer and Havens, 1990, and Witlox, 1994, given later in Figure 5.5). Since the data show that the dense gas effects gradually become more and more important as Ri_0 increases from about 1.0 to about 100, the "critical Ri_0" of 50 is an approximation that could easily be shifted by about a factor of two.

The following specific definitions of Ri_0 can be given in terms of known initial parameters:

Continuous plumes at ground level $\quad Ri_0 = \dfrac{g(\rho_{po} - \rho_a)}{\rho_a} \dfrac{V_{co}}{D_0 \, u_*^3}$ (5-1)

Instantaneous cloud at ground level $\quad Ri_0 = \dfrac{g(\rho_{po} - \rho_a)}{\rho_a} \dfrac{V_{io}}{D_0^2 \, u_*^2}$ (5-2)

where $(\rho_{po} - \rho_a)$ is the difference between the initial plume density and the ambient density, V_{co} is the initial volume flow rate for continuous plumes, V_{io} is the initial volume of an instantaneous cloud, D_0 is the initial cloud width, and u_* is the friction velocity (equal to about 5% to 10% of the wind speed at a height of 10 m).

To decide whether the dense gas cloud should be considered to be a "continuous" release [i.e., using equation (5-1)], or as an "instantaneous" release [i.e., using equation (5-2)], it is necessary to compare the duration of the release, T_d, with the travel time from the source to the receptor position, x/u. It is assumed that the wind speed and direction are fairly steady over the time period. When $T_d > x/u$, the cloud is likely to be continuous. When $T_d < x/u$, the cloud is likely to be instantaneous, since it "looks" like a puff (i.e., by the time the leading edge of the cloud reaches x, the trailing edge has detached itself from the source). For $T_d \sim x/u$, it is recommended that both the instantaneous and continuous formulas be applied, and the minimum concentration chosen (Britter and McQuaid, 1988).

The Ri_o criteria given above are appropriate for ground-based clouds. However, many source release scenarios involve elevated jets of dense gases from a "stack" height of h_s. If the initial cloud is dense enough, its trajectory will curve downward as it "sinks" toward the ground. The importance of the density effects on the initial trajectory curvature near the stack can be determined by substituting the initial diameter, D_p, of the rupture or pipe break for the cloud length scale, D_o, in equations (5-1) and (5-2):

Continuous elevated plumes (near source)

$$Ri_o = \frac{g(\rho_{po} - \rho_a)V_{co}}{\rho_a \, u_*^3 \, D_p} \tag{5-3a}$$

Instantaneous elevated plumes (near source)

$$Ri_o = \frac{g(\rho_{po} - \rho_a)V_{io}}{\rho_a \, u_*^2 \, D_p^2} \tag{5-3b}$$

The above definitions refer to dense gas effects on the plume or cloud near the source before it touches the ground. If one is also interested in whether the dense gas effects are influencing the maximum ground level concentrations after the dense plume sinks to the ground, then the stack height, h_s, should be used as the length scale in the Ri_o definitions:

Continuous elevated plumes (maximum ground level concentrations)

$$Ri_o = \frac{g(\rho_{po} - \rho_a)V_{co}}{\rho_a \, u_*^3 \, h_s} \tag{5-4a}$$

Instantaneous elevated cloud (maximum ground level concentration)

$$Ri_o = \frac{g(\rho_{po} - \rho_a)V_{io}}{\rho_a \, u_*^2 \, h_s^2} \tag{5-4b}$$

Very few laboratory and field experiments have been concerned with determining the critical Ri_o values for elevated clouds. Until new data are available, we assume that the $Ri_o = 50$ criterion developed for ground-level clouds also is valid for elevated clouds.

Note that the Ri_o value defined for the cloud near the source in equations (5-3) is larger than the Ri_o value defined for the ground level concentrations in equations (5-4) by a factor of h_s/D_p for continuous plumes and a factor of $(h_s/D_p)^2$ for instantaneous clouds, which can be as large as 100 or 1000. Thus it is possible that, for elevated sources, the trajectory of the cloud near the source (but at heights well above the ground) may be dominated by cloud density, while the behavior at distances where the maximum ground-level concentration occurs may be dominated by ambient turbulence.

It is noted that these Ri_0 criteria can all be expressed in terms of the buoyancy fluxes, $B_c = g(\rho_{po} - \rho_a) V_{co}/\rho_a\pi$ and $B_i = g(\rho_{po} - \rho_a) V_{io}/\rho_a\pi$, defined by Briggs (1995). These buoyancy fluxes are the "dense gas" equivalents of the buoyancy fluxes used in the stack plume rise formulas defined by Briggs (1984) for lighter-than-air plumes.

The Ri_0 criteria listed above should be tested prior to any model applications. Many of the modeling systems automatically test these or similar criteria and select appropriate modeling algorithms. For example, the EPA (1994) TSCREEN modeling system will apply the Britter and McQuaid (1988) dense gas modeling procedures if the initial cloud is sufficiently dense and will apply a standard EPA passive gas model if the initial cloud is not dense. The worked examples in Appendices A through G include calculations of Ri_0.

5.2. Jet Trajectory and Entrainment

In many accident scenarios, especially for pressurized releases, the vapor cloud is released as a jet with significant initial momentum. Hazardous gas concentrations can drop very rapidly in a jet, due to the enhanced entrainment of ambient air into the cloud caused by the large velocity differences between the jet and the ambient air. In contrast, gas concentrations in a low-momentum ground-based release may remain high much larger, due to much lower entrainment rates. Three types of specialized jet models are described below. The first type of model is for momentum jets from valve or pipe ruptures, the second type of model is for dense gas plumes from elevated releases, and the third type of model is for positively buoyant plumes from elevated releases.

5.2.1. Momentum-Dominated Jets

Many accidental releases from pressurized pipes and tanks are characterized by initial velocities, w_0, much greater than the local wind speed, u. In fact, as shown in Chapter 4, if the tank pressure exceeds about two times ambient pressure, these speeds are sonic, with magnitudes of 300 to 400 m s^{-1}. In these cases, the initial plume trajectory and dispersion are dominated by the jet velocity, and by the large entrainment rate, u_e. Briggs (1984) shows that the entrainment rate is proportional to the local jet speed, w. Typically, the concentration in a momentum jet can decrease by a factor of 100 over distances of less than 100 m. This situation can be compared with the variation of concentration in a low-momentum, ground-based dense-gas cloud where concentrations may decrease by less than a factor of two in the same distance.

Briggs (1984) describes the general theory of momentum and/or buoyancy-dominated plumes. For example, he suggests the following formula for the trajectory (height z_p versus distance x) of a plume which is initially directed upward. His formula is intended for positively buoyant plumes, but can also be expressed in a form applicable to negatively buoyant plumes:

$$z_p = \left(19 \frac{\rho_{po} M_o}{\rho_a u^2} x - 4.2 \frac{B_c}{u^3} x^2\right)^{1/3} \tag{5-5}$$

where z_p refers to height above the source, u is the wind speed at the source elevation, $M_o = w_o^2 D_o^2/4$ is the initial momentum flux and $B_c = g(\rho_{po} - \rho_a) w_o D_o^2/(4\rho_a)$ is the initial buoyancy flux, which we are assuming to be positive for dense plumes. Note that Briggs divides all "fluxes" by π. The first term (i.e., the momentum term) dominates equation (5-5) initially, until a distance $x = 2B_c u/M_o$ or a time $t = x/u = 2B_c/M_o$. Of course, when $z = -h_s$, the dense plume strikes the ground.

If the plume is momentum-dominated (i.e., if B_c can be assumed to be zero), the plume trajectory is given by the first term in equation (5-5). However, at some point the plume's internal motions are overwhelmed by the ambient turbulence and the plume stops moving across the wind. These crosswind trajectories of momentum jets that result from pressurized pipeline ruptures have been studied by Wilson (1979, 1981), who assumes that choked flow exists with sonic velocity at the pipe rupture. The plume travels a total distance, Δy, in a crosswind. The following equation is suggested by Wilson (1979) and is consistent with Briggs' (1984) formula for the maximum crosswind distance traveled by a momentum jet:

$$\Delta y = \frac{4.8 \rho_{po}^{1/2} M_o^{1/2}}{\rho_a^{1/2} u} \tag{5-6}$$

where the variables have been defined earlier. Note that there are no restrictions on whether the jet is pointed upward or downward; only that the jet must be perpendicular to the wind flow and there must be no obstructions to the jet. Thus the distance Δy could be in any direction perpendicular to the ambient wind direction. Field experiments in Alberta with typical sour gas pipeline ruptures directed upward show that plume travel, Δy, equals about 50 m, and that equation (5-6) is accurate within a factor of about two (Wilson, 1979).

These simplified formulas assume that there is no interaction of the jet with the ground. If the jet strikes the ground, it will be distorted laterally, although centerline concentrations may not change. Beyond a distance of about 100 m, the jet effects are rather unimportant for typical pipeline ruptures.

5.2.2. Elevated Dense Gas Jets

Special plume rise models are available to deal with continuous elevated releases of dense gas as defined by the Ri criteria in equations (5-3) and (5-4). These models also apply to neutrally buoyant momentum jets of the type discussed in Section 5.1.1. Bodurtha (1961), Ooms et al. (1974) and Hoot, Meroney and Peterka (1973) have pioneered in the study of this phenomenon, and Ooms and Duijm (1984) provide a useful review of the subject. A good example of the problem that must be modeled is seen in Figure 5-1, which contains temperature observations from a wind tunnel experiment by Xiao-Yun, Leijdens and Ooms (1986). The initial dense gas jet, which is released upward at the left side of the figure, rises at first due to its initial upward momentum from the stack, but then sinks due to its excess density. If the plume is dense enough, it may strike the ground surface, as seen in Experiments 2 and 3 in the figure.The available models for dense gas plume trajectories are all based on similar formulations. The Hoot et al. (1973) and Ooms et al. (1974) models are typical of this group of models, which include equations of mass, momentum, and energy conservation, and generally require integrations in time or distance. These dense gas jet models have been incorporated in more comprehensive dense gas models (e.g., the Ooms et al., 1974, model is now part of DEGADIS). All models require entrainment assumptions (regarding the rate at which the plume is diluted by ambient air) to close the system of equations and thus permit a solution. These models are also restricted to downwind distances of no more than a few hundred meters, where the internal turbulence in the plume dominates over the ambient turbulence. At greater distances, the excess density and/or momentum of the plume become insignificant compared to ambient conditions.

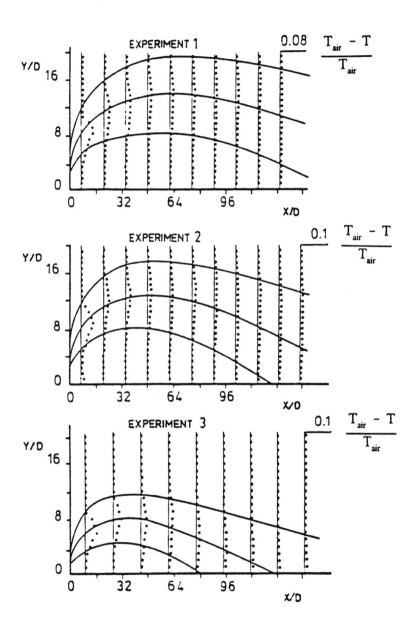

Figure 5-1. Observed (dots) relative temperature differences between plume and ambient condiions in a dense gas plume during three experiments in a wind tunnel. The effect of density is least in the top figure and greatest in the bottom figure. The three curved solid lines in each figure are predicted plume top, midpoint, and bottom. Y and X are vertical and downwind distance, and D is initial plume diameter. (After Xiao-Yun et al., 1986.)

Hoot et al. (1973) assume that the ambient cross-wind velocity, u, is constant in their dense gas plume model and that the distributions of variables within the plume are constant or uniform at any given distance from the source. Also, the specific heats of the plume and air are assumed equal. They then define the following basic equations, where the variables are defined in Figure 5-2.

$$\text{Entrainment: } \frac{d(R^2 u_s)}{ds} = a_1 R \, |u_s - u \cos \theta_p| + a_2 R u \, |\sin \theta_p| \quad (5\text{-}7)$$

(i.e., changes in volume flux are proportional to differences in velocities inside and outside of the plume around the boundary of the plume)

$$\text{Horizontal Momentum: } \frac{d(\rho_p R^2 u_s^2 \cos \theta_p)}{ds} = \rho_a u \frac{d(R^2 u_s)}{ds} \quad (5.8)$$

(i.e., the plume is accelerated by drag due to the momentum of the entrained air)

$$\text{Vertical Momentum: } \frac{d(\rho_p R^2 u_s^2 \sin \theta_p)}{ds} = (\rho_a - \rho_p) R^2 g \quad (5\text{-}9)$$

(i.e., vertical momentum changes are due to buoyancy differences)

$$\text{Mass: } \frac{d(\rho_p R^2 u_s)}{ds} = \rho_a \frac{d(R^2 u_s)}{ds} \quad (5\text{-}10)$$

(i.e, mass changes are due to entrainment of ambient air)

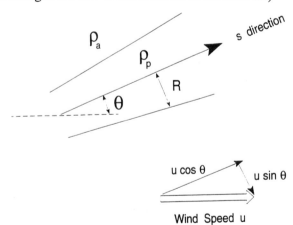

Figure 5-2. Drawing of plume parameters used in model of Hoot et al. (1973) corresponding to equations (5-7) through (5-10). The plume cross-section is circular with radius R.

The coordinate system is defined by s, which is the distance along the plume axis, and θ_p, which is the angle of this axis to the horizontal. R is plume radius, u_s is the plume speed in the direction s, and ρ_p is the plume density. The ambient atmosphere is defined by density ρ_a and speed u. The constants a_1 and a_2 in equation (5-7) are the constants for along-plume and cross-plume entrainment, which are found to equal 0.09 and 0.90, respectively, based on wind tunnel experiments. Equation (5-7) is an empirical entrainment assumption that is used to "close" the set of equations, or assure that the number of equations equals the number of variables. Equations (5-7) through (5-10) are solved analytically by Hoot et al. (1973) for an upward-pointing dense plume, leading to the following analytical expression for the maximum initial rise:

$$\frac{\Delta h}{2R_0} = 1.32 \left(\frac{w_0}{u}\right)^{1/3} \left(\frac{\rho_{po}}{\rho_a}\right)^{1/3} \left(\frac{w_0^2 \rho_{po}}{2R_0 g (\rho_{po} - \rho_a)}\right)^{1/3} \tag{5-11}$$

where w_0, ρ_{op}, and R_0 are the initial vertical plume speed, density and radius. The variables u and ρ_a are the ambient wind speed and density, respectively. This equation gives the height at which the dense plume stops rising and starts to bend down toward the ground.

The downwind distance, x_g, at which the centerline of the dense plume strikes the ground is given in the analytical relation:

$$\frac{x_g}{2R_0} = \frac{w_0 u \rho_{po}}{2R_0 g (\rho_{po} - \rho_a)} + 0.56 \left\{ \left(\frac{\Delta h}{2R_0}\right)^3 \cdot \left[\left(2 + \frac{h_s}{\Delta h}\right)^3 - 1\right] \cdot \frac{u^3 \rho_a}{2R_0 g w_0 (\rho_{po} - \rho_a)} \right\}^{1/2} \tag{5-12}$$

where $h_s/\Delta h$ is the ratio of stack height to maximum plume rise.

The ratio of the maximum concentration C to the initial concentration C_0 at two specific downwind positions is given by the analytical equations:

$$\frac{C}{C_0} = 1.688 \left(\frac{w_0}{u}\right) \left(\frac{\Delta h}{2R_0}\right)^{-1.85} \tag{5-13}$$

(at the point of maximum rise)

$$\frac{C}{C_0} = 2.43 \left(\frac{w_0}{u}\right) \left(\frac{h_s + 2\Delta h}{2R_0}\right)^{-1.95} \tag{5-14}$$

(at the point, x_g where the centerline strikes the ground)

The plume model by Ooms et al. (1974) differs from the previous model by Hoot et al. (1973) in a few ways. The basic conservation equations are treated in a slightly different way, and the cross-wind distribution of variables are assumed to have a Gaussian or normal shape in the Ooms model rather than a uniform shape. The closure or entrainment equation is assumed to include a term involving atmospheric turbulence:

$$\frac{d}{ds}\left(\int_0^{\sqrt{2}R} \rho_p u_s 2\pi r\, dr\right) = 2\pi R \rho_a\left(a_1\, |u_s - u\cos\theta_p| + a_2 u\, |\sin\theta_p|\cos\theta_p + (\varepsilon_d R)^{\frac{1}{3}}\right)$$

(5-15)

where the variables are defined as before, a_1 equals 0.057 and a_2 equals 0.50, and ε_d is the ambient eddy dissipation rate (which can be approximated by $u_*3/0.4z$ and typically equals about $10^{-3}m^2s^{-3}$). The plume boundary is assumed to be located at $\sqrt{2}R$ for the purposes of the integration. Except for the last term, this equation is similar to equation (5-7). The set of equations is solved by numerical integration given initial parameters.

Nearly all the vapor cloud models surveyed later in Chapter 7 contain a dense gas jet algorithm along the same lines as the models described above. The models are similar in the sense that they use formulas such as equations (5-7) through (5-10) and they have all "calibrated" their entrainment parameters with experiments such as those shown in Figure 5-2. Some models such as HGSYSTEM (Witlox, 1994) and PHAST (DNV Technica, 1993) also account for aerosols and plume thermodynamics. For example, as aerosols evaporate, the plume temperature would decrease.

Worked examples of dense gas jet releases are given in Scenario 1 (pressurized chlorine jet), Scenario 4 (normal butane jet), and Scenario 6 (multicomponent jet), presented in Appendices A, D, and F, respectively.

5.2.3. Positively Buoyant Plumes

In some cases, the plume from a hazardous gas release is positively buoyant (i.e., its density is *less* than the ambient density), allowing the use of standard buoyant plume rise models (Briggs, 1984). Methane and hydrogen are examples of chemicals whose initial buoyancy is positive. Several decades of research have gone into the development of plume rise models for typical buoyant industrial sources. EPA air quality models such as ISC (EPA, 1995), and TSCREEN (EPA, 1994), contain algorithms for continuous momentum plumes that have either neutral or positive buoyancy. A continuous release is one for which the release rate is constant over a time period exceeding the averaging time at the receptor, or for which the duration of the release exceeds the travel time from the source to receptor. These plume rise models are extensively reviewed in other documents (e.g., Hanna, Briggs, and Hosker, 1982, Briggs, 1984, or Turner, 1994) and will not be reviewed here. For example, equation (5-5) is an example of one of these general equations, as applied to calculate the trajectory of plumes from stacks. Once the initial buoyancy and momentum fluxes, B_o and M_o, are defined using the expressions following equation (5-5), then the plume trajectory, the final plume rise, and the dilution can be estimated. For buoyant plumes, B_o should be defined so it is positive upwards, giving a predicted "final plume rise" of $21.4 \, B_o^{3/4}/u$ at a downwind distance of $49 \, B_o^{5/8}$, according to the EPA's ISC model (EPA, 1992). An example of an application of this plume rise model and the ISC model is given in the hazardous gas Scenario 5 (warm H_2SO_4 plume) discussed in Appendix E.

5.3. Dense Gas Release at Grade

5.3.1. Background and Overview

The models for jets or momentum-dominated releases described in Section 5.2 are relevant for gas jets that are positively, negatively or neutrally buoyant. Many accidental releases, however, are from low-momentum continuous or quasi-instantaneous sources near the ground surface. Furthermore, the hazardous gas/aerosol mixture is often dense. In this case, the momentum-dominated models in Section 5.2 are not applicable and a new set of models is required. In this section, so-called box or slab models for dense gas releases at grade are reviewed, which assume constant concentrations of plume material at any downwind cross-section. Similarity models are also included in this discussion, since they merely assume a similar form for the concentration distribution at any cross-section. For example the Gaussian model is a similarity model, since it always assumes that the crosswind distribution has the same shape.

If the density, ρ_p, of the cloud is very close to the density of air, then the cloud may be transported and dispersed as if it were passively or neutrally buoyant (see Section 5.1). In order to determine if the excess density of the cloud is of concern, the initial cloud Richardson number should be calculated, using the definitions in equations (5-1) through (5-4).

The "critical Ri" should not be thought of as an absolute number, which separates dense gas-dominated from passive gas-dominated dispersion. Instead, as seen later in Figure 5-5, it should be interpreted as an approximation that only provides guidance about where these two effects may be roughly separated. In reality, dense gas effects gradually become important as Ri_0 increases, and the physical processes are slightly different depending on the source configuration.

Up until about 1970, the dispersion of dense gas clouds was treated in the same way as the dispersion of inert neutrally buoyant clouds. Van Ulden's (1974) experiments with dense gas clouds convinced everyone that something different should be done, since his observed plume lateral dispersion parameters, σ_y, were a factor of four larger than those used for neutrally buoyant clouds. His observed vertical dispersion parameters, σ_z, were a factor of four less than those for neutrally buoyant clouds. This tendency is called dense gas "slumping," and is similar to the flow of a spill of molasses on a table top. The problem of dense gas slumping has received much attention in basic and applied research programs over the past 20 years. There are several excellent review papers on the subject of dense gas slumping (e.g., Raj, 1985; Fay, 1986; Britter and McQuaid, 1988; and Britter, 1989).

If the sole problem were the slumping of a dense gas cloud at ambient temperature, then the problem could be greatly simplified. However, it has been observed that some gases with molecular weights less than air (e.g., ammonia) can behave like dense gases due to their cold temperatures and/or the presence of aerosols. Because many dense gas clouds are cold and contain aerosols, the following thermodynamic phenomena may need to be modeled for a particular scenario:

- Latent heat exchanges due to phase changes (evaporation and condensation) in the cloud.
- Heat exchanges with the underlying surface.
- Exothermic (heat releasing) and endothermic (heat absorbing) chemical reactions.

Figure 5-3 characterizes these effects on a schematic diagram. Note that the complicated thermodynamic aspects relative to the source emissions from the gas/aerosol/liquid spill are not treated here because they have been covered separately in Section 4. The operational models surveyed in Section 7 and evaluated in Section 8 generally account for, at most, only one or two of the phenomena in Figure 5-3. The modelers have checked the magnitudes of all these terms and concluded that most of them, such as heat losses from and to opaque clouds, are insignificant over the lifetime of a typical hazardous cloud. However, opaque clouds can mask the underlying liquid pool, thus reducing the solar energy reaching the pool surface.

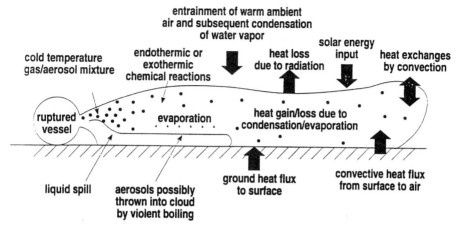

Figure 5-3. Thermodynamic aspects of a typical hazardous material release.

5.3.2. Dense Gas Clouds in the Absence of Heat Exchange

In order to simplify the discussion, it is useful to focus on the dispersion of the dense gas cloud in the absence of heat exchanges. This simplified system has been the subject of many research programs and field and laboratory experiments. It is important to first point out that many gases with molecular weights less than that of air (e.g., ammonia, hydrogen fluoride) will act like dense gases when they are accidentally released to the atmosphere, because of the frequent presence of aerosols in the cloud. If the aerosol drops occupy a small volume fraction of the total mixture, then the density of a multicomponent two-phase cloud is given by the relation

$$\frac{1}{\rho_p} = \frac{1}{p_a/RT_p} + \frac{1}{\rho_{aer}} \tag{5-16}$$

where ρ_{aer} is the mass of aerosol per unit cloud volume (kg m^{-3}), p_a is the atmospheric pressure (in newtons m^{-2}), T_p is the absolute temperature (°K) of the plume or cloud, and R, the "gas constant" for the gas mixture, is given by the formula:

$$R = R^* \sum_{j=1}^{n} \frac{m_j}{m} \frac{1}{M_j} \tag{5-17}$$

where R^* is the universal gas constant (8.31436 joules mole^{-1} °K^{-1}), M_j is the molecular weight of gas component, j, m is the total mass of the gases in the mixture, and m_j is the mass of gas component j. The term p_a/RT_p in equation (5-16) represents the density of the gas phase of the mixture. For the purposes of the derivations below, assume that the total mass of aerosol does not change (i.e., there are no phase changes and no liquid rain-out).

First consider an instantaneous dense gas release with initial volume, V_{io}, initial height h_o, and initial radius, R_o. If the density perturbation in the cloud is significant, then the cloud will have a tendency to spread along the ground in the horizontal direction rather than diffusing in the vertical direction. The horizontal speed of the edge of the instantaneous cloud, dR/dt, at any time after release will be proportional to the square root of the cloud potential energy at that time:

$$\frac{dR}{dt} \propto \left(g \frac{\rho_p - \rho_a}{\rho_a h} \right)^{\!\!1/2} \propto \left(g \frac{\rho_p - \rho_a}{\rho_a} \frac{V_i}{\pi R^2} \right)^{\!\!1/2} \tag{5-18}$$

where it is assumed that the cloud continues to have a cylindrical shape, with height h, radius R and volume V_i, over its lifetime. The constant of proportionality is found to equal about unity. If there are no heat additions to the cloud, then the buoyancy term, ($g[(\rho_p - \rho_a)/\rho_a]V_i$) for the instantaneous cloud in equation (5-18) will remain constant (Fay, 1986) and the following well-known formula for the cloud radius will result:

$$R^2 = R_o^2 + 2 \left(g \frac{\rho_p - \rho_a}{\rho_a} V_i \right)_{\!o}^{\!1/2} t \tag{5-19}$$

where subscript "o" indicates the initial value.

This formula has been validated with both field and laboratory data, as seen by the observations from the Thorney Island field experiments given in Figure 5-4. The source was an instantaneously released volume (about 2000 m^3) of Freon gas. Except for the "flow establishment" zone at small times, the data clearly justify the use of equation (5-19). Note that the radius R of the slumping cylinder defined here is horizontally oriented, whereas the radius R of the elevated dense gas plume defined in Figure 5-2 is perpendicular to the plume axis.

The simple box or slab model approach (assuming uniform concentrations across the cloud) to the dense gas slumping problem is seen to work well, even though videotapes of observed slumping clouds show a "doughnut" shape to the cloud. The advancing dense gas cloud is clearly marked

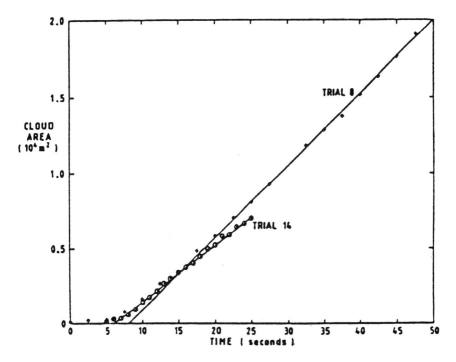

Figure 5-4. Area of dense gas cloud versus time for two Thorney Island trials (Brighton et al., 1985).

in these video tapes by a turbulent frontal zone. However, these detail in the flow structure and their associated concentration gradient have little influence on the gross cloud behavior, which is determined by integrated parameters such as the buoyancy flux, $g[(\rho_p - \rho_a)/\rho_a)V_i]$.

The derivations above were applicable to an instantaneous dense gas release. In the case of a continuous dense gas release, slumping in the along-wind direction can usually be ignored compared to advection by the wind. The increased spread is primarily felt in the lateral or cross-wind direction, as defined by the lateral cloud radius, R. Raj (1985) rewrites equation (5-18) in a form appropriate for continuous plumes:

$$u_a \frac{dR}{dx} \propto (g\frac{\rho_p - \rho_a}{\rho_a}h)^{\frac{1}{2}}$$

$$(5\text{-}20)$$

To solve this equation the effective advection velocity of the plume, u_a, must be known. This velocity increases as the plume grows vertically and represents the wind velocity at some concentration-weighted mean height.

Some researchers (e.g., Witlox, 1994a,b) assume that the effective wind speed is equal to the wind speed at some fraction (say 0.4) of the plume depth. It could also be assumed that the effective wind speed is equal to the observed wind speed at some height times the term $[1 - (V_{co}/V_c)]$, where V_{co} and V_c are the initial and current volume fluxes. In this case the plume accelerates simply because it entrains ambient air. It is simplest for a model to assume that u_a equals the wind speed at some standard measurement height, say 2 m or 10 m (Britter and McQuaid, 1988, assume 10 m), although this speed will be an overestimate when the plume is shallow and an underestimate when the plume is deep. If there are not heat additions to the plume, the buoyancy flux, $B_c = 2gu_ahR(\rho_p - \rho_a)/\rho_a$, will be conserved, yielding the following solution (Raj, 1985) for a continuous dense gas plume:

$$R = R_o \left\{ 1 + 1.5 \left[gh_o \frac{(\rho_p - \rho_a)_o}{\rho_a} \right]^{1/2} \frac{x}{u_aR_o} \right\}^{2/3} \qquad (5\text{-}21)$$

Note that the cloud width grows at a faster rate for the continuous plume than for the instantaneous cloud, since the instantaneous cloud is free to spread in the along-wind direction.

Equations (5-19) and (5-21) provide solutions for the rate of growth of the radius of the cloud resulting from instantaneous and continuous dense gas releases. Because time of travel equals x/u_a, it is implied that, as the ambient wind speed decreases, the dense gas cloud growth in the *upwind* direction may exceed the wind speed, allowing the cloud to spread upwind. Eventually, as it becomes more dilute, the cloud will stop advecting upwind. Britter (1980) studied laboratory data and suggested that the dense cloud would advect upwind a distance Δx from the source center given by the formula:

$$\Delta x_{upwind} = R_o + 2 \left(g \frac{\rho_p - \rho_a}{\rho_a} V_c \right)_o u_*^{-3} \qquad (5\text{-}22)$$

where the expression $(g[(\rho_p - \rho_a)/\rho_a]V_c)_o$ is the initial buoyancy flux, B_c. Britter also finds that, as a result of the same physical principles causing the upwind spreading phenomenon, the dense cloud will spread laterally by a width $\Delta y_{initial}$ at the location of the source:

$$\Delta y_{initial} = 2R_o + 8 \left(g \frac{\rho_o - \rho_a}{\rho_a} V_c \right)_o u_*^{-3} \qquad (5\text{-}23)$$

It is seen that the cube of the friction velocity, u_*, appears in equations (5-22) and (5-23), implying that accurate knowledge of u_* is crucial to the estimation of upwind and cross-wind dense gas spread at light wind speeds. A factor of two uncertainty in u_* leads to a factor of eight uncertainty in Δx_{upwind} and $\Delta y_{initial}$.

The emphasis in the above derivations is on the horizontal plume spread, R. The vertical dimension of the plume, h, is also important and can be estimated from the above equations if the rate of growth of the continuous plume volume flow rate or the instantaneous cloud volume are known. As seen in the Thorney Island experiments, if the cloud is dense enough, h can decrease with time near the source. The growth in plume volume causes a decrease in concentrations due to dilution, as given by the formulas:

$$\text{Instantaneous} \qquad \frac{C}{C_o} = \frac{V_i}{V_i} = \frac{h_o R_o{}^2}{h R^2} \qquad (5\text{-}24)$$

$$\text{Continuous} \qquad \frac{C}{C_o} = \frac{V_{co}}{V_c} = \frac{q_o}{u_a h R} \qquad (5\text{-}25)$$

where C_o is the initial cloud concentration and V_{io} and V_{co} are the initial instantaneous volume and the initial continuous volume flux, respectively. This time rate of increase in cloud volume is parametrized by use of the entrainment velocity, v, defined by the following equations:

$$\text{Instantaneous} \qquad \frac{1}{V_i}\frac{dV_i}{dt} = \frac{v}{R} \qquad (5\text{-}26)$$

$$\text{Continuous} \qquad \frac{1}{V_c}\frac{dV_c}{dt} = \frac{v}{R} \qquad (5\text{-}27)$$

where R is a representative plume distance scale. The gross entrainment velocity v, in equations (5-26) and (5-27) is usually broken down into two components—one, v_{top}, that describes entrainment through the top of the cloud, and another, v_{edge}, that describes entrainment through the edge of the cloud.

With this breakdown, equations (5-26) and (5-27) can be written as follows for an instantaneously released cylindrical cloud of height h and radius R and for a continuously released cloud with cross-section given by height h and radius R:

$$\text{Instantaneous} \quad \frac{1}{V_i}\frac{dV_i}{dt} = \frac{v_{\text{edge}}}{2\pi R} + \frac{v_{\text{top}}}{h} \tag{5-28a}$$

$$\text{Continuous} \quad \frac{1}{V_c}\frac{dV_c}{dt} = \frac{v_{\text{edge}}}{2R} + \frac{v_{\text{top}}}{h} \tag{5-28b}$$

$V_c = 2u_aRh$ and $\dot{V}_i = \pi R^2 h$ for these "slab" geometries.

Some earlier dense gas models (e.g., Van Ulden, 1974; Eidsvik, 1980) accounted for the edge entrainment, which they thought to be dominant. Now, most models assume that top entrainment is dominant and assume the following functional form for the top entrainment rate, which is defined as $u_e = v_{\text{top}}$:

$$u_e = bu_* f_e(Ri_*) \tag{5-29}$$

where u_* is the friction velocity (a measure of ambient turbulence), b is a proportionality constant, and f_e is a universal function of the local cloud Richardson number Ri_*, generally defined as

$$Ri_* = g\frac{\rho_p - \rho_a}{\rho_a}\frac{h}{u_*^2} \tag{5-30}$$

Note that this definition is different from that for Ri_o in equations (5-1) through (5-4). Initially, $Ri_* = (u_*/u_a)Ri_o$ for continuous plumes and $Ri_* = (u_*/u_a)^2 Ri_o$ for instantaneous clouds. Furthermore, the emphasis of equation (5-30) is on local ρ_p and h at any downwind distance, x. The "constant" b in equation (5-29) is in the range from 0.4 to 0.8, consistent with the known growth of passive plumes (Briggs, 1984), for which $Ri_* = 0$. Most estimates of u_e, b, and f_e are based on laboratory experiments with continuous plumes from line sources, so that the effects of cross-wind (lateral) spread are eliminated, and equation (5-28b) reduces to $u_e = dh/dt$. Thus, observations of the growth of the cloud top can be used to infer the entrainment rate, u_e.

The formulas for u_e used in two commonly applied dense gas models, DEGADIS (Spicer and Havens, 1990) and HEGADAS (McFarlane et al., 1990; Witlox, 1994a) are plotted along with laboratory observations in Figures 5-5a and 5-5b, respectively. It is noted that the laboratory data on which these curves are based are mostly about 20 years old. In the Ri_* range from about 1 to 100, both models base their formulas on the so-called

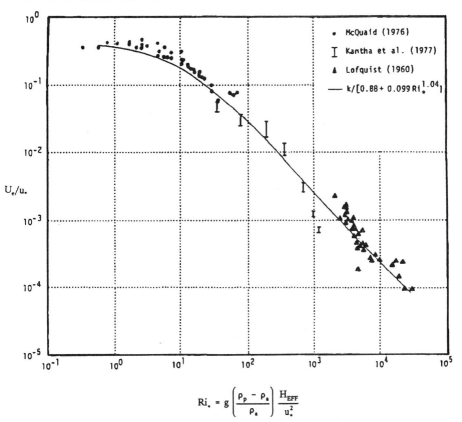

$$Ri_* = g\left(\frac{\rho_p - \rho_a}{\rho_a}\right)\frac{H_{EFF}}{u_*^2}$$

Figure 5-5a. Dimensionless entrainment velocity u_e/u_* as a function of the bulk Richardson number, Ri_*, used in DEGADIS (Spicer and Havens, 1990). $Ri_* = (u_*/u)Ri_0$, where Ri_0 is used in equation (5-1). The solid line represents the DEGADIS curve: $u_e/u_* = 0.4/(0.88 + 0.099\ Ri_*^{1.04})$. Data from McQuaid (1976), Kantha et al. (1977), and Lofquist (1960) are plotted.

McQuaid (1976) data, which are available only from in-house data summaries. Clearly there is a need for new, comprehensive, research programs to better define the entrainment rate, u_e.

The DEGADIS and HEGADAS formulas for u_e agree within ±10 or 20% at Ri_* less than about 10, but disagree at very large Ri_* due to their different slopes:

$$\text{DEGADIS} \quad \frac{u_e}{u_*} = \frac{0.4}{0.88 + 0.099 Ri_*^{1.04}} \tag{5-31}$$

$$\text{HEGADAS} \quad \frac{u_e}{u_*} = \frac{0.6}{(1.0 + 0.8 Ri_*)^{\frac{1}{2}}} \tag{5-32}$$

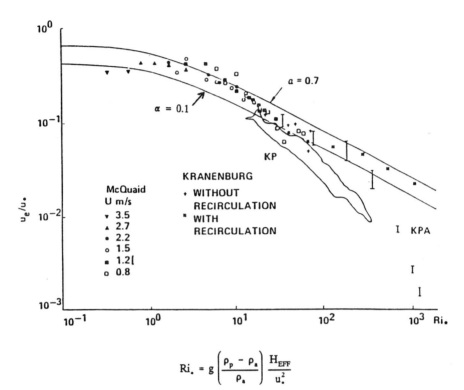

$$Ri_* = g \left(\frac{\rho_p - \rho_a}{\rho_a} \right) \frac{H_{EFF}}{u_*^2}$$

Figure 5-5b. Dimensionless entrainment velocity u_e/u_* as a function of the bulk Richard-son number, Ri_*, used in HEGADAS (McFarlane et al., 1990). $Ri_* = (u_*/u)Ri_o$, where Ri_o is used in equation (5-1). The error bars indicate the results of the Kantha et al. (1977) (KPA) experiments, corrected for side-wall drag. The Kato and Phillips (1969) (KP) data are continuously distributed within the cloud outlined in this Figure. The solid lines represent McFarlane et al.'s (1990) fit to the Kranenburg (1984) and McQuaid (1976) data: $u_e/u_* = 0.41 (1 + \alpha)/\phi(Ri_o)$, where separate curves are drawn for $\alpha = 0.1$ (low z_0, neutral) and $\alpha = 0.7$ (high z_0, stable), and $\phi(Ri_*) = (1.0 + 0.8\ Ri_*)^{1/2}$.

The ratio of the DEGADIS to HEGADAS u_e at $Ri_* > 10$ is equal to about $6Ri_*^{-0.54}$. Thus, at $Ri_* = 10^3$, the DEGADIS u_e is only 14% of the HEGADAS u_e. However, Britter (1989) points out that the laboratory data at $Ri_* > 100$ are largely irrelevant to dense gas clouds in the real atmosphere, since such very dense gas clouds would quickly slump, causing their depth, h, to be greatly reduced. Most real atmosphere problems will then never have to deal with the large Ri_* end of Figures 5.5a and 5.5b.

The entrainment data in Figures 5.5a and 5.5b are used by all dense gas modelers to derive their own forms of equation (5-29). As a result, our reviews of many of the models discussed in Chapter 7 reveal little differences in the predictions of entrainment rates, u_e, in the region where $Ri_* < 100$.

It is interesting to note that the variation of the cloud volume ratio, V_i/V_{io}, (assumed equal to the concentration ratio, C_o/C) with downwind distance for instantaneous releases during the Thorney Island experiments (McQuaid, 1985) showed little dependence on initial cloud Richardson number, as shown in Figure 5-6. It should be noted that this Richardson number is defined in the same way as in equation (5-30). C is the maximum concentration observed by monitors within the cloud as it passed a given monitoring arc. Even though the excess cloud density affects the cloud radius, R, and depth, h, individually, there is evidently little effect on the cloud volume, $Q = \pi h R^2$. Any decreases in cloud height, h, were compensated by increase in cloud area, πR^2. The initial volume, Q_o, in these trials is about 2000 m^3 in all cases. Instantaneous releases of Freon gas took place in the experiments shown on the figure, with wind speeds ranging from about 1 ms^{-1} to 8 ms^{-1} and $(\rho_p - \rho_a)/\rho_a$ ranging from about 1.4 to 4. The line on the figure is given by the formula:

$$\frac{V_i}{V_{io}} = \left(\frac{x}{V_{io}^{1/3}}\right)^{1.5} \quad \text{valid for } x \geq V_{io}^{1/3} \tag{5-33}$$

The scatter of the points on the figure is relatively small as far as atmospheric data are concerned, and are all within a factor of two of the line given by equation (5-33). It is noted, however, that since the initial volume, V_{io} remained near 2000 m^3 for all Thorney Island experiments, equation (5-33) has not been tested over a range of V_{io}.

Hanna et al. (1993) also found a lack of dependence of maximum cloud concentration at a given downwind distance, x, on Ri_0 for field experiments involving continuous dense gas clouds. This phenomenon can be explained, as above, by the fact that, even though the cloud height, h, may be smaller for very dense plumes, the cloud radius, R, increases such that the product, hR, and hence the centerline concentration, does not change.

As shown by Briggs (1995), the dense cloud depth, h, can be calculated directly from the estimates given above for the growth of cloud radius, R, and volume flux, V_c, or volume, V_i. The rate of increase of h is different for different entrainment relations between u_e and Ri_*, as illustrated by Briggs' (1995) results for a continuous dense gas plume:

$$\text{If } u_e \propto Ri_*^{-1} \text{ (as in DEGADIS)}, \quad h = A_1 u_*^3 u^{-1} B_0^{-2/3} x^{5/3} \tag{5-34}$$

$$\text{If } u_e \propto Ri_*^{-1} \text{ (as in HEGADAS)}, \quad h = A_2 u_*^2 u^{-1} B_0^{-1/3} x^{4/3} \tag{5-35}$$

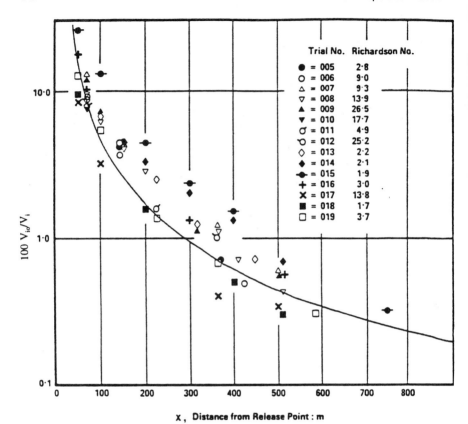

Figure 5-6. Thorney Island field observations of dilution as a function of a downwind distance. V_i is defined as the cloud volume, and it is assumed that $V_{io}/V_i = C/C_o$. The predicted curve from equation (5-33) is drawn. McQuaid (1985) presented the original figure minus the predicted curve. Richardson number is defined by equation (5-30).

where the "constants" A_1 and A_2 equal about 2.0 and 1.0, respectively. The greater-than-linear growth of h with x may seem counter-intuitive, since neutral or passive plumes are known to have a linear (or less) growth of h with x. This result can be explained by considering that the dense gas cloud starts out with a smaller h, and then accelerates up to passive growth as the cloud dilutes.

The dense gas dispersion formulas given above are all based on the presumption that the surface is a homogeneous dirt-covered area or grassy field and that the effects of nearby hills and ditches can be ignored. All field experiments have been carried out under these simplified conditions (for example, the Thorney Island experiments took place on an abandoned airfield in England). However, most accidental releases are likely to occur in industrial sites characterized by numerous buildings, tanks, pipes, etc., and surrounded by a mixture of houses, commercial areas, and other industries. The characteristic roughness length, z_0, of the field experiment sites is about 0.001 or 0.01 m, and that of industrial sites is about 1 m.

The increased roughness length over an industrial site affects the predicted concentration by enhancing the friction velocity, u_*, which then enhances the entrainment rate, u_e [see equation (5-29)], and by reducing the wind speed near the ground. The increase in u_* is generally found to be the dominant effect. Equation (3-4) suggests that, for neutral conditions and for the assumption of constant wind speed at a height, z, the ratio of friction velocities over surfaces 1 and 2 is given by:

$$\frac{u_{*2}}{u_{*1}} = \frac{\ln (z/z_{01})}{\ln (z/z_{02})} \tag{5-36}$$

If the wind speed can be assumed to be constant at an elevation of 100 m and if $z_{01} = 0.01$ m and $z_{02} = 1.0$ m, then equation (5-36) leads to the result that the ratio u_{*2}/u_{*1} is about two. On the basis of a theoretical analysis, Hanna (1992) suggested the general rule of thumb that for dense gas releases, predicted concentrations decrease by about 20% to 50% for each factor of ten increase in surface roughness. The results of sensitivity studies with computer models such as HEGADAS and SLAB confirm this rough estimate.

The effects of terrain are ignored in most operational hazardous gas models based on box or slab assumptions under the rationale that the primary impacts of accidental releases will occur near the source, before the cloud encounters a hill or a ditch. However, since dense gas spread is initially gravity driven, it is clear that there will be a downhill-directed force (proportional to the size of the slope angle) on the dense cloud. DeNevers (1984) and Kukkonen and Nikomo (1992) have studied the problem for simple slopes. For example, DeNevers (1984) compared the downslope flow speed component to the gravity-spreading speed for a typical dense gas release and found that they became equal for slopes as small as 1 or 2 degrees. Thus, for a uniform slope of 2 degrees, the dense gas cloud modeled by DeNevers (1984) would be moving downslope faster than it was spreading. The effects of uniform slopes and other types of terrain complications can be directly simulated by three-dimensional Eulerian models such as FEM3C, as described by Chan (1994) (see Section 5.5).

5.3.3. Dense Gas Clouds in the Presence of Heat Exchanges

The derivations in section 5.3.2 were for a dense gas cloud that did not exchange heat with its environment or generate heat internally through latent heat processes or chemical conversions. In practical situations, the atmosphere will attempt to eliminate any temperature or density gradients that are initially present. For example, a cold cloud resulting from a spill from a pressurized tank will be warmed by heat transfer from the underlying ground or sea surface, by entrainment of warm ambient air, and by latent heat released by condensation of ambient water vapor. On the other hand it could be cooled by the emission of long wave radiation from the upper surface of its aerosol cloud or by evaporation of aerosol droplets. These internal heat changes can also affect the dispersion of the cloud, since they modify the cloud density. In any given problem, some of these heat exchange terms are more significant, and in another problem, others may be more significant. The model developers have analyzed the magnitudes of the various heat exchange terms and retained only those that are significant. Methods of parametrizing many of these heat transfer parameters are discussed below:

Heat transfer to the ground or water surface from below: The rate at which heat can be added to the cold cloud from below is limited by the thermal conductivity and hence the heat flux within the substrate. This is the same problem addressed in Section 4.2.5 on liquid pool evaporation. Obviously water is a more efficient conductor of heat than loose, dry gravel, although a complication can occur with water if the surface freezes and the conductivity decreases. The temperature at a given level in the ground or water will continually decrease as heat is drawn to the surface. The energy equation can be simplified such that the local rate of change of temperature with time is given by the Fourier heat conduction law:

$$\frac{\partial T}{\partial t} = \frac{k_s}{c_{ps}\rho_s} \frac{\partial^2 T}{\partial z^2} \tag{5-37}$$

where k_s, c_{ps}, and ρ_s are the thermal conductivity (W m^{-1} K^{-1}), specific heat (J kg^{-1} K^{-1}), and density (kg m^{-3}) of the soil or water. Boundary conditions are that T approaches a constant at great depths, and that the flux— $k_s\partial T/\partial z$ approaches the sensible plus latent heat fluxes in the air from the ground to the cloud (in the absence of radiative effects). The thermal conductivity k_s is observed to range over several orders of magnitude, and values for several typical surfaces are given by Oke (1978). For example, the Shell SPILLS models uses values of 0.32 and 2.21 W m^{-1} K^{-1} for dry and wet soil, respectively.

Heat transfer from the surface to the cloud: The primary heat transfer mechanism from the surface to the cloud is often the sensible heat flux, H_s (W m^{-2}). If the temperature difference between the ground and the cloud is large (say 10°C or greater) then heat transfer is by free convection (i.e., independent of the wind speed). For such conditions, Wu and Schroy (1979) suggest the empirical formula:

$$H_s = h_c(T_s - T_{air})$$

$$(5\text{-}38)$$

where T_s is the ground surface temperature, T_{air} is the cloud temperature, and h_c is an empirical convection coefficient, assumed equal to 4.5 W m^{-2} K^{-1}. For smaller temperature differences (less than 10°C), the wind speed must be accounted for and a different empirical formula is used.

Heat transfer by evaporation/condensation processes within the cloud: If the cloud initially contains an aerosol with density ρ_{aer}, then that aerosol will take from the cloud an amount of heat (per unit volume of the cloud) equal to $H_v\rho_{aer}$ when it all evaporates. H_v is the latent heat of evaporation for the chemical. The rate of evaporation depends on the temperature variation and the vapor pressure of the chemical within the dense gas cloud. This mechanism is responsible for the maintenance of the density difference in aerosol clouds whose molecular weight is less than that of air (e.g., ammonia). The evaporation of aerosols is said to have a "refrigerator" effect.

As ambient air is entrained by the cold, dense cloud, the ambient air is cooled. Depending on the water vapor mixing ratio in the ambient air and the temperature of the mixture, pure water drops can be condensed from the entrained air. As these water drops condense they release their latent heat of vaporization to the cloud. The maximum amount released per unit volume is equal to the latent heat of vaporization for water times the difference between the saturation density of water vapor in the cloud and the actual density of water vapor in the ambient air. These evaporation/condensation formulas are likely to be overestimates, since a feedback mechanism is operating whereby the heat exchanged during the phase change alters the temperature of the cloud.

Temperature change by dilution with ambient air: The ambient atmosphere and the slumping cloud are nearly always turbulent and thus are very diffusive. By the time a cloud of hazardous materials is transported 1 km downwind, its volume will have increased by a large amount due to entrainment of ambient air. Assuming no other heat inputs and equal molar heat capacities, the new temperature of the mixture is given by the formula:

$$\rho V_i T \text{ (mixture)} = \rho_a(V_i - V_{io})T_{air} + \rho_o V_{io} T_o \qquad (5\text{-}39)$$

where ρ and V_i are the molar density and the volume of the mixture, and subscript "o" refers to the initial gas release. This equation strongly influences the heat exchange processes within the cloud because of the temperature change of the mixture. Methods of estimating the dilution V_i/V_{io} are reviewed at several places in this section (e.g., see Figure 5-6 and equation 5-33).

It is noted that worked examples of the dispersion of dense gases released at ground-level with low initial momentum are given in Appendix B (Scenario 2—liquid spill of refrigerated chlorine), Appendix C (Scenario 3—acetone spill), and Appendix G (Scenario 7—area source of HF).

5.4. Transport and Dispersion of Neutrally Buoyant or Passive Gas Clouds

Application of the Ri_o criterion in Section 5.1 may suggest that the cloud is likely to behave more like a neutrally buoyant or passive gas cloud than at dense gas cloud. Or, even the most dense gas clouds will eventually no longer be strongly influenced by dense gas effects at large travel times or downwind distances, where the cloud density approaches the ambient density. Some entrainment models have been written for simplified box or slab models that include interpolated expressions that account for density effects and for entrainment due to ambient turbulence (e.g., Eidsvik, 1980, or Ermak, 1990). However, most of the operational models surveyed in Chapter 7 and evaluated in Section 8 employ arbitrary assumptions marking the transition point between the dense gas and passive gas algorithms. The most commonly used transition assumption is that dense gas effects cease when the density perturbation $(\rho_p - \rho_a)/\rho_a$ drops below some limit (e.g., 0.01 or 0.001). For example the DEGADIS code uses a limit of 0.001.

Beyond the transition point, most box or slab models such as SLAB, DEGADIS, and HGSYSTEM change over to a Gaussian plume or puff model (Hanna et al., 1982; Turner, 1994) which expresses the ground-level concentration for a neutrally buoyant or passive plume using the following formulas:

$$\text{Plume: } \frac{C}{Q} = \frac{1}{\pi\sigma_y\sigma_z u}\exp\left[-\frac{(y-y_o)^2}{2\sigma_y^2}\right]\exp\left[-\frac{h_p^2}{2\sigma_z^2}\right] \tag{5-40}$$

$$\text{Puff: } \frac{C}{Q_i} = \frac{1}{\sqrt{2}\pi^{3/2}\sigma_x\sigma_y\sigma_z}\exp\left[-\frac{(y-y_o)^2}{2\sigma_y^2}\right]\exp\left[-\frac{h_p^2}{2\sigma_z^2}\right]\exp\left[-\frac{(x-x_o)^2}{2\sigma_x^2}\right]$$
$$\tag{5-41}$$

where Q is the mass emission rate (kg s^{-1}) for a continuous plume and Q_i is the mass (kg) in the initial instantaneous puff. h_p is the elevation of the cloud centerline (midpoint), $(y - y_o)$ is the crosswind distance from the cloud axis, and σ_y and σ_z are the standard deviations of the lateral and vertical distributions of concentration. Table 5-1 lists the σ_y and σ_z formulas recommended by Hanna et al. (1982). For a puff, the along-wind position $(x - x_o)$ and along-wind dispersion σ_x must also be considered, as seen in equation (5-41). The distance x_o can be assumed to equal $u \cdot t$, where t is the time after release of the puff and u is the average advection speed. Note that the name "Gaussian" refers to the normal or Gaussian shape of the distribution curve.

The transition from a dense gas slab model, with cloud depth h and cloud half-width R, to a Gaussian model can be accomplished by means of a virtual source procedure, where for a puff it can be assumed that the dispersion parameters σ_z, σ_y, and σ_x are given by the relations:

$$\sigma_z = \frac{h}{\sqrt{2}} \tag{5-42}$$

$$\sigma_y = \frac{R}{\sqrt{2}} \tag{5-43}$$

$$\sigma_x = \sigma_y \quad \text{(if instantaneous source or puff)} \tag{5-44}$$

It is assumed that an imaginary or a "virtual" source exists some distance x_v upstream from the transition point, such that the maximum concentration in the Gaussian plume would equal the value calculated by the box or slab model. The virtual source distance could be larger or smaller than the actual distance, x, depending on the conditions. The lateral and vertical dispersion parameters, σ_y and σ_z, in the Gaussian dispersion model can be written as functions of x or t and inverted to solve for the virtual source distance, x_v, or time, t_v. For example, if $\sigma_y = 0.1x$, then $x_v = 10R/\sqrt{2}$. The equations in Table 5-1 can be used directly in the Gaussian model to calculate virtual distances.

TABLE 5-1
Formulas Recommended by Hanna, Briggs, and Hosker (1982) for $\sigma_y(x)$ and $\sigma_z(x)$(10 m < x < 10 km) for Continuous Plumes in Rural and Urban Terrain

Pasquill-Gifford Stability Class	σ_y (m)	σ_z(m)
Rural Conditions		
A	$0.22x(1 + 0.0001x)^{-1/2}$	$0.20x$
B	$0.16x(1 + 0.0001x)^{-1/2}$	$0.12x$
C	$0.11x(1 + 0.0001x)^{-1/2}$	$0.08x(1 + 0.0002x)^{-1/2}$
D	$0.08x(1 + 0.0001x)^{-1/2}$	$0.06x(1 + 0.0015x)^{-1/2}$
E	$0.06x(1 + 0.0001x)^{-1/2}$	$0.03x(1 + 0.0003x)^{-1}$
F	$0.04x(1 + 0.0001x)^{-1/2}$	$0.016x(1 + 0.0003x)^{-1}$
Urban Conditions		
A–B	$0.32x(1 + 0.0004x)^{-1/2}$	$0.24x(1 + 0.001x)^{1/2}$
C	$0.22x(1 + 0.0004x)^{-1/2}$	$0.20x$
D	$0.16x(1 + 0.0004x)^{-1/2}$	$0.14x(1 + 0.003x)^{-1/2}$
E–F	$0.11x(1 + 0.0004x)^{-1/2}$	$0.08x(1 + 0.0015x)^{-1/2}$

The dispersion parameters for instantaneous sources (duration of release less than averaging time or less than travel time from source to receptor, with a maximum time limit imposed by the time over which the wind is steady) are known to be different than those for plumes from continuous sources (Hanna et al., 1982). But since there are many more data available for continuous sources, most models use the continuous source dispersion parameters for all types of sources. As discussed in Section 3.3, stability class A is very unstable, D is neutral, and F is very stable. The dispersion rates decrease in general as atmospheric stability increases. Tables 5-2 and 5-3 provide a mechanism for estimating the stability class, which depends on observations of windspeed and cloudiness, and knowledge of the sun's elevation angle. Hanna et al. (1982) discuss alternate methods of estimating stability class based on vertical temperature gradients (a method favored by the Nuclear Regulatory Commission) or lateral turbulence intensity (a method sometimes used by the EPA if onsite data are available).

TABLE 5-2
Meteorological Conditions Defining Pasquill-Gifford Stability Classes
(Gifford, 1976)
Insolation Category is Determined from Table 5-3.

A: Extremely unstable conditions
B: Moderately unstable conditions
C: Slightly unstable conditions

D: Neutral conditions
E: Slightly stable conditions
F: Moderately stable conditions

	Daytime insolation			Nighttime Conditions		Anytime
Surface wind Speed, m/sec	Strong	Moderate	Slight	Thin overcast or >4/8 low cloud	≥3/8 cloudiness	Heavy Overcast
<2	A	A–B	B	F	F	D
2–3	A–B	B	C	E	F	D
3–4	B	B–C	C	D	E	D
4–6	C	C–D	D	D	D	D
>6	C	D	D	D	D	D

TABLE 5-3
Method of Estimating Insolation Category, Where Degree of Cloudiness Is
Defined as That Fraction of the Sky above the Local Apparent Horizon That Is
Covered by Clouds

Degree of Cloudiness	Solar Elevation Angle >60°	Solar Elevation Angle ≤60° but >35°	Solar Elevation Angle ≤35° but >15°
4/8 or less or any amount of high thin clouds	Strong	Slight	Slight
5/8 to 7/8 middle clouds (2000 m to 5000 m base)	Moderate	Slight	Slight
5/8 to 7/8 Low Clouds (less than 2000 m base)	Slight	Slight	Slight

In the case of instantaneous puff releases, some models use σ_x, σ_y, and σ_z formulas derived directly from field experiments involving puff releases, as summarized by Slade (p. 175 of *Meteorological and Atomic Energy*, 1968) and listed in Table 5-4:

TABLE 5-4
Dispersion Formulas for Instantaneous Releases (from Slade, 1968)

Stability Class	A	B	C	D	E	F
σ_y or σ_x	$0.18x^{0.92}$	$0.14x^{0.92}$	$0.10x^{0.92}$	$0.06x^{0.92}$	$0.04x^{0.92}$	$0.02x^{0.89}$
σ_z	$0.60x^{0.75}$	$0.53x^{0.73}$	$0.34x^{0.71}$	$0.15x^{0.70}$	$0.10x^{0.65}$	$0.05x^{0.61}$

Slade gave formulas only for stability classes B, D, and F, which we interpolated and extrapolated to stability classes A, C, and E in the table. The formulas in Table 5-4 are based on far fewer observations than the continuous plume formulas in Table 5-1. In particular, σ_x observations are generally lacking, partly because of difficulties in interpretation of the effects of wind shears on σ_x.

A Gaussian model for passive gas releases can be developed from the equations given above. Or, the user can obtain a copy of the EPA's (1995) Industrial Source Complex (ISC) software and user's guide, which includes similar equations. The ISC model has wide recognition by regulatory agencies.

The simple Gaussian model for passive gases is applied in Appendix E for Scenario 5, which is concerned with the release of H_2SO_4 from a building vent.

5.5. Simple Nomograms for Calculating the Dilution of Dense Gas Release

As stated earlier, it is possible to derive useful scaling relations by applying physical insights to the dense gas dispersion problem. As an example of this approach, Britter and McQuaid (1988) suggest a set of simple but useful equations and nomograms in their *Workbook on the Dispersion of Dense Gases*. They collected the results of many laboratory and field studies of dense gas dispersion, plotted the data in dimensionless form, and drew curves or nomograms that best fit the data. These relations, which are given below, are best suited to instantaneous or continuous ground level area or volume sources of dense gases.

The following physical variables and parameters are used in the nomograms:

C_m (kg m^{-3})   Concentration (mean) across cloud

C_0 (kg m^{-3})   Initial concentration

V_{io} (m^3)	Initial cloud volume (for instantaneous puff releases)
V_{co} (m^3 s^{-1})	Initial plume volume flux (for continuous plume releases)
u (m s^{-1})	Wind speed at $z = 10$ m
T_d (s)	Duration of release
x (m)	Downwind distance
ρ_o (kg m^{-3})	Initial cloud density
ρ_a (kg m^{-3})	Ambient gas density
g_o' (m s^{-2}) $= g\,(\rho_o - \rho_a)/\rho_a$	Initial "reduced gravity" term
D_i (m) $= V_{io}^{1/3}$	Characteristic source dimension for instantaneous release
D_c (m) $= (V_{co}/u)^{1/2}$	Characteristic source dimension for continuous release

Other variables such as surface roughness length, averaging time, and atmospheric stability class are not included in this list because the authors conclude that available data for dense gas releases do not show any strong influence of these parameters. Also, the effects of the initial source (e.g., jet, aerosol) are assumed to be unimportant at the downwind distances of interest. Consistent with the sets of field and laboratory data used in the analysis, it can be concluded that the representative averaging time, T_a, for the continuous plumes in these experiments is about 3 to 10 minutes, the representative roughness length is a few centimeters (that is, a flat grassy surface), and the representative stability class is about C or D (that is, neutral to slightly unstable).

The variables in the above list were combined by Britter and McQuaid (1988) into dimensionless variables and used to develop the following physical relations:

$$\text{Continuous Releases:} \qquad \frac{C_m}{C_o} = f_c\left(\frac{x}{(V_{co}/u)^{1/2}},\ \frac{g_o' V_{co}^{1/2}}{u^{5/2}}\right) \qquad (5\text{-}45)$$

$$\text{Instantaneous Releases:} \qquad \frac{C_m}{C_o} = f_i\left(\frac{x}{V_{io}^{1/3}},\ \frac{g_o' V_{io}^{1/3}}{u^2}\right) \qquad (5\text{-}46)$$

where f_c and f_i refer to universal dimensionless functions. It is important to recall that u is defined by Britter and McQuaid (1988) as the observed value at a height of 10 m. The first term on the right hand side of equations (5-45) and (5-46) is a dimensionless distance, and the second term is a source Richardson number, which is defined differently than the Ri_o in equations (5-1) through (5-4). For example, equations (5-45) and (5-46) use a representative wind speed, u, rather than the friction velocity, u_*. Britter and McQuaid (1988) plotted various sets of laboratory and field observations of C_m/C_o on graphs whose vertical axis is the dimensionless distance and whose horizontal axis is the Richardson number, and found that the data were indeed well-organized by the relations suggested by equations (5-45) and (5-46). The authors then drew contours of constant C_m/C_o, resulting in the nomograms presented as Figures 5-7 and 5-8.

The dashed boxes labeled "full-scale data region" indicate ranges of data covered by the field observations. The vertical line labeled "passive limit" marks values of Ri below which the cloud loses its dense-gas behavior.

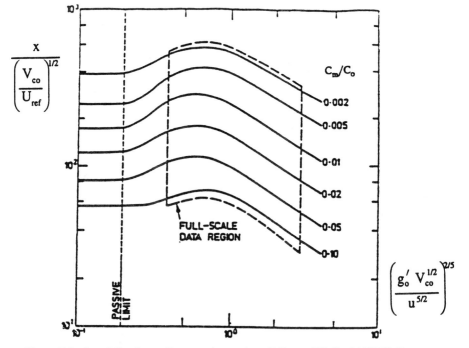

Figure 5-7. Correlation for continuous releases from Britter and McQuaid (1988). Parameters are defined in and above equation (5-45):

$$\frac{C_m}{C_o} = f_c\left(\frac{x}{(V_{co}/u)^{1/2}}, \frac{g_o' V_{co}^{1/2}}{u^{5/2}}\right)$$

The passive gas C_m/C_o limits are intended to be approximations to the Gaussian plume model for neutral conditions (Class D).

A decision must be made whether a given release should be considered to be continuous or instantaneous (i.e., Figure 5-7 or 5-8). A criterion has been developed that is based on the ratio of the plume length, uT_d, to the downwind distance, x, as expressed by the following conditions:

$uT_d/x \geq 2.5$ $\rightarrow$ Continuous

$uT_d/x \leq 0.6$ $\rightarrow$ Instantaneous

$0.6 \leq uT_d/x \leq 2.5$ $\rightarrow$ Calculate both ways, take minimum predicted concentration

Computer software has been written containing analytical equations that represent, by linear segments, the C_m/C_o curves in the two nomograms

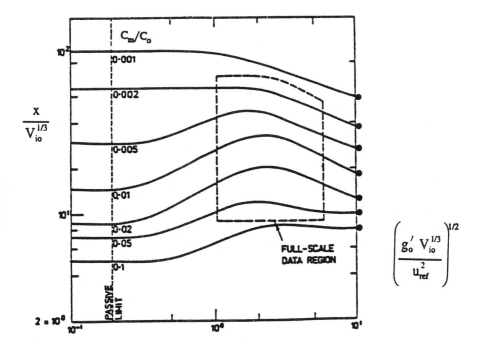

Figure 5-8. Correlation for instantaneous release from Britter and McQuaid (1988). Parameters are defined in and above equation (5-46):

$$\frac{C_m}{C_o} = f_i \left(\frac{x}{V_{io}^{1/3}} , \frac{g_o' V_{io}^{1/3}}{u^2} \right)$$

presented in Figures 5-7 and 5-8. These equations were used in the model evaluations reported in Chapter 8 and have been incorporated in the USEPA's TSCREEN model (EPA, 1994). In order to assure that C_m/C_o smoothly approaches 1.0 as x approaches 0.0, the analytical equations include the following interpolation formulas at small x (that is, for $x \leq 30D_c$ or $x \leq 3D_i$, where D_c and D_i are the initial source dimensions for continuous and instantaneous releases):

$$\frac{C_m}{C_o} = \frac{306(x/D_c)^{-2}}{1 + 306(x/D_c)^{-2}} \quad \text{Continuous Plume} \quad (5\text{-}47)$$

$$\frac{C_m}{C_o} = \frac{3.24(x/D_i)^{-2}}{1 + 3.24(x/D_i)^{-2}} \quad \text{Instantaneous Cloud} \quad (5\text{-}48)$$

As Britter and McQuaid (1988) point out, their nomograms are not intended to solve all dense gas problems at all distances over all types of terrain and ambient conditions, but are intended to provide guidance that incorporates the primary physical principles at plant fenceline distances, where the major impacts are expected to occur. The method becomes less appropriate for the near-source region of jets or for two-phase plumes. However, the authors point out that the jet effect is usually minor at downwind distances beyond about 100 m. Furthermore, they suggest a method for accounting for the effects of a two-phase ammonia cloud, and illustrate the procedure for an ammonia accident. With this method, it is assumed that enough ambient air is mixed into the ammonia plume to completely evaporate the unflashed liquid and that the initial density equals the air–ammonia mixture density at the normal boiling point of ammonia ($T = 240°K$). Once the initial density, ρ_o, and volume, Q_o, are calculated, it is assumed that there are no thermodynamic processes acting in the cloud (that is, $T_{cloud} = T_{air}$) in subsequent calculations.

We emphasize that the Britter and McQuaid model is suggested for use only as a benchmark screening model. It should not be applied to scenarios outside of its range of derivation. For example, it may be inappropriate for application in urban or industrial areas, where the surface roughness is large. Other more detailed models such as SLAB and DEGADIS are sufficiently general that they allow for variations in input parameters such as surface roughness and averaging time.

5.6. Three-Dimensional Numerical Models of Dense Gas Dispersion

The simple box, slab, or similarity models discussed in Sections 5.1 through 5.4 can provide great insight into the physical processes important for the transport and dispersion of hazardous materials. Furthermore, they are shown to agree fairly well with the limited data bases (see Chapter 8). Yet it is obvious that there is much structure within the plume or cloud that is not being accounted for by the models. Furthermore, if the cloud is influenced by a nearby terrain feature or building, the flow and concentration patterns are altered. A few advanced numerical models have been developed that do account for varying gradients and flux divergences within the plume. Most of these are based on solutions to the three-dimensional, time-dependent conservation equations, and often require a mainframe supercomputer. For example, a set of governing equations used by the FEM3C model developed by Chan (1994) is listed below:

Equations of Motion

$$\frac{\partial(\rho_p \mathbf{u})}{\partial t} + \rho_p \mathbf{u} \cdot \nabla \mathbf{u} = -\nabla(p - p_a) + \nabla \cdot (\rho_p K^m \cdot \nabla \mathbf{u}) + (\rho_p - \rho_a)\mathbf{g} \quad (5\text{-}49)$$

Equation of Mass Continuity

$$\nabla \cdot (\rho_p \mathbf{u}) = \frac{\partial \rho_p}{\partial t} \quad (5\text{-}50)$$

Equation of State

$$\rho_p = \frac{pM}{R^* T} = \frac{p}{R^* T \left((C''/M_j) + [(1 - C'')/M_a] \right)} \quad (5\text{-}51)$$

Mass Conservation of Pollutant Vapor

$$\frac{\partial(\rho_p C'')}{\partial t} + \mathbf{u} \cdot \nabla \rho_p C'' = \frac{1}{\rho_p} \nabla \cdot (\rho K^c \cdot \nabla \rho_p C'') + \left(\frac{\partial(\rho_p C'')}{\partial t} \right)_{pc} \quad (5\text{-}52)$$

Mass Conservation of Pollutant Liquid

$$\frac{\partial(\rho_p C_l'')}{\partial t} + \mathbf{u} \cdot \nabla \rho_p C_l'' = \frac{1}{\rho_p} \nabla \cdot (\rho K^c \cdot \nabla \rho_p C_l'') + \left(\frac{\partial(\rho_p C'')}{\partial t} \right)_{pc} \quad (5\text{-}53)$$

Conservation of Enthalpy

$$\frac{\partial \theta}{\partial t} + \mathbf{u} \cdot \nabla \theta = \frac{1}{\rho_p c_p} \nabla \cdot (\rho_p c_p \mathsf{K}^\theta \cdot \nabla \theta) + \frac{c_{pn} - c_{pa}}{c_p}(\mathsf{K}^c \cdot \nabla \rho_p C'') \cdot \nabla \theta - \frac{L}{c_p}\left(\frac{\partial(\rho_p C'')}{\partial t}\right)_{pc}$$

$$(5\text{-}54)$$

where vector (bold) and tensor (san serif) notation is used and the dot represents the dot product (e.g., $\mathbf{u} \cdot \nabla C'' = u\partial C''/\partial x + v\partial C''/\partial y + w\partial C''/\partial z$). The following definitions are used:

ρ_p = Total plume density (kg m^{-3})

C'' = Concentration of hazardous gas or vapor in the plume (kg kg^{-1})

C_1'' = Concentration of hazardous liquid in the plume (kg kg^{-1})

c_p = Specific heat of the mixture (J kg^{-1} K^{-1})

c_{pn} = Specific heat of the hazardous gas (J kg^{-1} K^{-1})

c_{pa} = Specific heat of the air (J kg^{-1} K^{-1})

∇ = Del operator ($\mathbf{i}\partial/\partial x + \mathbf{j}\partial/\partial y + \mathbf{k}\partial/\partial z$)

$\mathbf{u}$ = Wind vector ($\mathbf{i}u + \mathbf{j}v + \mathbf{k}w$)

M_j = Molecular weight of the hazardous gas

M_a = Molecular weight of air

K = Eddy diffusivity coefficient (m^{2}s^{-1}), which is a function of local turbulence and stability

pc = phase change (subscript)

p = Pressure in the plume (newtons m^{-2})

p_a = Atmospheric pressure in the absence of the plume (newtons m^{-2})

Superscripts on the Ks refer to the values appropriate to the variable in question.

Note that phase changes (e.g., evaporation or condensation of droplets) within the cloud are accounted for. The above set of equations is solved by computer over a three-dimensional set of grid points, given appropriate initial and boundary conditions. Any set of equations describing dispersion also requires a closure hypothesis. The box or slab models use an entrainment assumption to close the system. Similarly, the set of equations (5-49) through (5-54) require specification of the K coefficients to close the system. These coefficients are not well known for some situations (e.g., very stable conditions) important for hazardous gas systems and probably are the major contributors to uncertainties in the three dimensional numerical models.

Chan et al. (1987) use the following formulas for the K^m's (note that only the diagonal terms of the tensor K are assumed to be nonzero):

$$\text{Vertical } K^m_z = 0.4[(u_* z)^2 + (w_* l)^2]^{1/2}\, \varphi^{-1}(Ri) \tag{5-55}$$

$$\text{Horizontal } K^m_y = 0.4 b u_* z \varphi^{-1}(Ri) \tag{5-56}$$

where

$$\varphi(Ri) = 1 + 5Ri \qquad\qquad (Ri \geq 0) \tag{5-57}$$

$$\varphi(Ri) = (1 - 16Ri)^{-1/4} \text{ for momentum} \qquad (Ri < 0) \tag{5-58}$$

$$\varphi(Ri) = (1 - 16Ri)^{-1/2} \text{ for energy and species} \quad (Ri < 0) \tag{5-59}$$

and

$$Ri = \frac{u_*^2 Ri_a}{u_*^2 + w_*^2} + 0.05 \frac{\rho_p - \rho_a}{\rho_p} \frac{gl}{u_*^2 + w_*^2} \tag{5-60}$$

where g is an adjustable parameter and l is the scale length of the cloud (equal approximately to the cloud depth). The convective velocity w_* is proportional to the cube root of the ambient heat flux times the mixing depth. The local Richardson number, Ri, defined in equation (5-60), is seen to be a function of the ambient air Richardson number, Ri_a, and a term involving the internal stratification of the dense gas cloud. The ambient Richardson number is defined by the wind speed and temperature gradients at height z: $Ri_{air} = (g/T_{air})\,(\partial T_{air}/\partial z + 0.01\ K/m)/(\partial u/\partial z)^2$.

The assumptions regarding the vertical dispersion coefficient, K^m_z, are arbitrary but reasonable, and satisfy the requirement that the effective scale of the turbulence be less than the scale of the cloud. But the form of the equation for the horizontal dispersion coefficient, K^m_y, would lead to much discussion among modelers. Many modelers prefer to use a K^m_y formulation such that K^m_y is independent of height.

Comparisons of FEM-3 predictions with observations are given in Figures 5-9 and 5-10. Peak concentrations patterns for Thorney Island Trial 17 are seen to be fairly accurately simulated in Figure 5-9, taken from Chan et al. (1987). The observed concentration time series from the Falcon field tests in Figure 5-10 are seen to be more variable than the model predictions, although the model is able to simulate the maximum concentrations and gross time variability within a few percent.

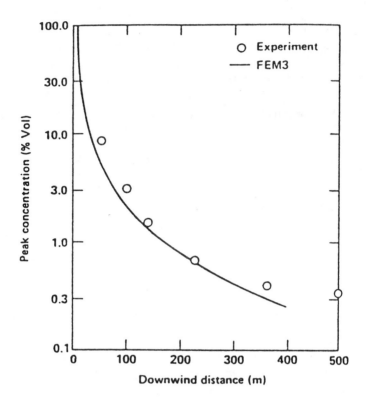

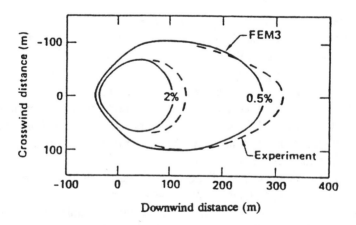

Figure 5-9. FEM-3 predicted versus measured peak concentrations (left) and contours (right) for Thorney Island Trial No. 17 (at height = 0.4 m), from Chan et al. (1987).

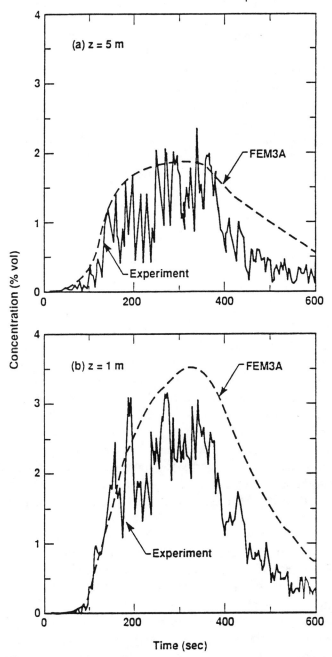

Figure 5-10. Falcon-4 FEM-3 predicted versus measured concentration for two different heights behind a fence, from Chan (1994).

An advantage of numerical models is that they can more precisely treat sources which have time and space variability in their source characteristics. For example, the emissions from individual grids covering an evaporating pool can be given as a function of time. Drawbacks are the large amounts of computer time and storage required, and uncertainties introduced by numerical approximations. Other models of this type have been proposed by Deaves (1984), England et al. (1978), and Riou and Saab (1985). These models can produce voluminous output information, which needs to be carefully edited for communications with users.

5.7. Transport and Dispersion Near Buildings

Releases of hazardous gases are likely to occur in the vicinity of buildings, which may modify the boundary layer wind flow patterns so that the trajectory and rate of dilution of the plume may be altered. In the past, nearly all hazardous gas models have ignored the influence of buildings and other obstacles because the model developers believed that these obstacles generally lead to enhanced dilution and lower concentrations. However, several modelers have recently included the effects of buildings, because there are certain circumstances when concentrations increase in the presence of buildings. For example, a plume could be constrained by the walls of buildings, or a plume from a short stack could be mixed to the ground in the wake of a building, leading to increased ground level concentrations.

The algorithms described below are based on the results of reviews of the effects of structures on toxic vapor dispersion carried out by Schulman et al. (1990), Britter (1989), and Brighton (1989). Meroney (1987) prepared a comprehensive set of guidelines for wind tunnel modeling to address this problem. Sufficient information exists to develop models for a few simplified source–receptor combinations. In the case of some source scenarios, the problem is so complex that it is necessary to carry out further wind tunnel or field studies. Fortunately, experimental data and empirical formulas have been developed for the three primary topics described below (i.e., lateral confinement in building canyons, concentration patterns on building faces due to vent releases, and downwash into building wakes). However, nearly all of these experiments and formulas are most valid for passive (i.e., neutrally buoyant) gases and relatively little information is available for dense gas dispersion around buildings.

5.7.1. Plume Confinement by Canyons

Consider the scenario in which the hazardous gas source is near or within a canyon between large buildings. The building height is H_B and the canyon width is W_C. Laboratory experiments by Konig (1987) and Marotske (1988) suggest that maximum concentrations are increased by a factor of as much as three due to the confinement by the canyon. This effect can be decreased if the plume height, h, grows so that leakage occurs above the buildings. If the source is between the buildings or if the source is upwind of the buildings and $\sigma_y < W_C/\sqrt{12}$ when the plume enters the canyon, and if $h/H_B < 1$, then a Gaussian plume model can be used in which σ_y is not allowed to exceed the limit:

$$\sigma_y \text{ (maximum)} = \frac{W_C}{\sqrt{12}} \tag{5-61}$$

This value of σ_y corresponds to a uniform lateral distribution across W_C. If $h/H_B > 1$, then only the lower part of the plume is confined by the canyon, and the upper part of the plume is free to disperse laterally as if the canyon were not there. In this case, the following interpolation formula can be used:

$$\text{Effective } \sigma_y \text{ (maximum)} = \frac{H_B}{h}\frac{W_C}{\sqrt{12}} + \frac{h - H_B}{h} \cdot \sigma_y \text{ (without canyon)} \tag{5-62}$$

where σ_y (without canyon) refers to the lateral dispersion as ordinarily calculated by the Gaussian plume model in the absence of obstacles. When the plume reaches the end of the canyon, lateral diffusion resumes, and a virtual source procedure can be applied to calculate concentrations further downwind.

5.7.2. Concentrations on Building Faces Due to Releases from Vents

If a hazardous gas is released accidentally within a building, it is likely to be exhausted by vents that typically take the form of flush vents on the roof or the sides of the building or very short stacks on the roof of the building. The concentration in the plume in the exhaust vent will have been affected by dilution through the volume inside the building. Because there is usually minimal buoyancy to the plumes being vented, the gas can be modeled as if it were neutral or passive. As seen in the reviews by Meroney (1982) and

Wilson and Britter (1982), there have been many wind tunnel studies of distributions of dimensionless concentration,

$$C^* = \frac{u_H A C}{Q} \tag{5-63}$$

on the faces of buildings of various shapes due to releases from vents on various positions on the buildings. A is a representative area of the building, assumed to equal $H_B W_B$, and u_H is the wind speed at the height of the building in the flow upwind of the building. The source and receptors are assumed to be on the same or adjacent faces. Using the definitions in Figure 5-11, the maximum concentrations on the building at a distance, r, from the source, are given by the formulas:

$C = 9(Q/u_H r^2)$ for $r/A^{1/2} < 1.73$ and source and receptor
                        on upper 2/3 of building   (5-64)

$C = 30(Q/u_H r^2)$ for $r/A^{1/2} < 1.73$ and source and receptor
                        on lower 1/3 of building   (5-65)

The restriction that $r/A^{1/2} < 1.73$ in equations (5-64) and (5-65) is applied so that there is a smooth transition from these formulas to the formula in the next section for the near wake (i.e., $C = 3Q/u_H A$). As distance, r,

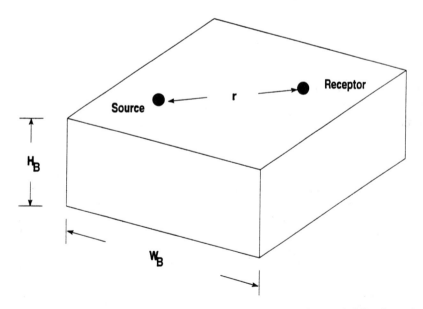

Figure 5-11. Definitions of parameters for calculating concentrations on building faces due to emissions from vents.

decreases, the concentration does not increase indefinitely but is capped by the concentration in the vent exhaust. These formulas say nothing about the lateral or vertical extent of the plumes from the vents. Equations (5-64) and (5-65) are most useful for estimating maximum concentrations with the condition that the plume is being blown directly from the source vent to the receptor position on the building face. Note that if there is a significant air flow from the vent, the vent plume may be transported up and away from the roof. In this case, the concentrations given by equations (5-64) and (5-65) would be conservative.

5.7.3. Concentrations on the Building Downwind Face (the Near-Wake) Due to Releases from Sources on the Building

An algorithm can be derived that provides a smooth transition from the vent scenario covered above. The source emissions are again assumed to be neutral or passive, and we are now concerned with the concentration in the near-wake or the recirculating cavity. This is a turbulent well-mixed zone that extends about two to five building dimensions downwind, and it is assumed that concentrations are uniform across this zone. Wilson and Britter (1982) find that, when the pollutants are well-mixed across the wake, the concentrations in the near wake are given by

$$C = \frac{3Q}{u_H A}\left(\frac{r}{A^{\frac{1}{2}}} \geq 1.73\right) \tag{5-66}$$

where $A = W_B H_B$ for blockish buildings and $A = H_B^{4/3} W_B^{2/3}$ for wide buildings, with a limit of $W_B = 8H_B$ for very wide buildings. Figure 5-12 provides a schematic depiction of this scenario. Note that the condition $r/A^{1/2} \geq 1.73$ is applied to equation (5-66), where r is the distance from the source to the receptor.

5.7.4. Other Effects of Buildings

The three simplified algorithms to account for the effects of buildings covered in the sections above tend to increase concentration impacts at ground level. Other building effects have been ignored here because they tend to have little effect on concentrations or to significantly decrease concentrations. The "flat-terrain no building" solution is conservative. For example, if a source is *upwind* of a building, fence, or other obstacle, the increased turbulence due to the obstacle will tend to dilute the hazardous gas plume. The effects of fences have been investigated in field and

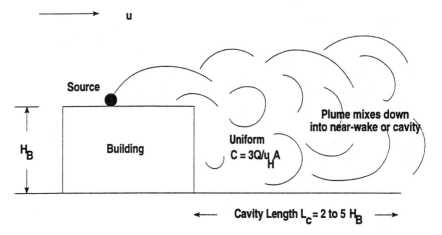

Figure 5-12. Schematic diagram of passive plume mixing into near-wake.

laboratory experiments because of their potential for mitigating the plume, and enhanced dilutions of a factor of three or more have been observed. Also, shallow dense gas plumes approaching taller obstacles are seen to be caught in the horseshoe vortices formed around the obstacle, and are therefore transported laterally away from the obstacle. Wind tunnel experiments are sometimes concerned with the transport and dispersion around special obstacle types, such as ramps (Britter and Snyder, 1988), urban areas (Schatzman, 1995), and uniform arrays of square blocks (Heidorn et al., 1992).

If appropriate, most hazardous gas models could be modified to include the EPA's downwash algorithm for the *far* wake as implemented in their Industrial Source Complex (ISC) model (EPA, 1992). This algorithm applies to the scenario when there is a stack of significant height (h_s equal to about 1 to 2 H_B) near the building and ground-level concentrations are to be calculated at a distance of about 10 H_B or greater from the stack. The algorithm allows for enhancement of σ_y and σ_z, depending on the ratio h_s/H_B. This algorithm is not implemented in most hazardous gas models at the present time because most sources are at the ground or at vents on the roofs of buildings, and maximum impacts would occur in the near wake during the more likely release scenarios. The ISC model and its building wake algorithm are applied in Appendix E to Scenario 5 (H_2SO_4 release from vent on building).

5.8. Worst Case Meteorological Conditions

When vapor cloud models are run in "planning mode," it is customary to identify a few worst-case scenarios that would have the strongest influence on control or mitigation strategies. The worst case is generally defined with respect to maximum ground-level concentrations at receptors on the boundaries of the industrial plant property or beyond. For near-ground-level continuous releases of passive (i.e., neutrally buoyant) or dense gases, the worst case is associated with stable conditions and light winds, for which dispersion and dilution are minimized. This conclusion is valid only for continuous releases.

For passive gases released continuously or instantaneously from stacks or vents at heights of about 50 m or greater, the worst-case scenario may not be stable conditions, since vertical dispersion may be so slight that the bottom of the plume does not reach the ground. Hanna et al (1982) show that, for passive gas releases at heights of about 50 m to 100 m, the worst-case ground-level impacts are associated with neutral conditions with moderate wind speeds. The situation becomes more complicated for buoyant plumes (e.g., power plant plumes), whose plume rise is inversely proportional to windspeed. Experience has shown that, for strongly buoyant plumes emitted from 100- 200-m-tall stacks, maximum ground-level concentrations occur during strongly convective conditions (i.e., sunny summer days).

If a dense gas release near the ground is instantaneous or of short duration (i.e., the time of release is less than the time of travel to the receptor), the worst-case conditions are likely to be associated with moderate wind speeds. Short-time releases of dense gases tend to spread out widely near the source when winds are light, thus delaying their advection downwind. Because the winds are light, by the time the cloud reaches a receptor at a distance of 500 m or 1000 m, it is relatively dilute. In contrast, when winds are higher, the dense cloud does not spread out widely near this source but is advected downwind with less dilution. This situation may also occur for slightly elevated short-duration dense gas releases, if they sink to the ground within a few tens of meters of the stack. Because the worst case for instantaneous or short-duration dense gas releases is not well-defined, it is advisable to carry out model runs for a range of stabilities and wind speeds for each case.

For instantaneous passive (neutrally buoyant) releases near the ground, there is no initial dense gas slumping, and the worst-case is likely to be associated with stable conditions and light winds.

5.9. Removal by Dry and Wet Deposition

A hazardous plume may consist of a mixture of gases, solid particles, and aerosols. Larger particles and aerosols will fall to the ground due to gravitational settling. Smaller particles and aerosols and gases will deposit on the surface due to a process called dry deposition, which is caused by a combination of phenomena such as chemical reactions and physical interception by the ground and vegetation (Sehmel, 1984). In the presence of rain, fog, or snow, the pollutant may be removed from the plume and deposited on the ground either by absorption or collection by the water drops or snow flakes (Slinn, 1984).

Models for hazardous gas dispersion (e.g., SLAB, DEGADIS, or HGSYSTEM) generally do not account for dry or wet deposition, since, at distances within a few hundred meters of the source, these processes are significant only for very large particles (diameter $D_p > 1000$ μm). Most of the research on dry and wet deposition has been connected with much larger time and space scales (e.g., acid deposition over the northeastern United States over time periods of several days). In addition, there has been concern about deposition of toxic substances such as dioxin which are produced at solid waste incinerators. Comprehensive wet and dry deposition modules have been built into the EPA's Regional Acid Deposition Model (RADM) and Industrial Source Complex-Version 3 (ISC-3) model.

5.9.1. Gravitational Settling of Large Particles or Aerosols

Large particles or aerosols, with diameter, D_p, greater than about 50 μm (where 1 μm = 10^{-6} m), will have a gravitational settling speed, v_s, of greater than 0.1 m/s. In this case, removal is dominated by simple gravitational settling. Stoke's law can be used (EPA, 1995) to calculate the gravitational settling speed of particles with diameters of 1 μm or larger:

$$v_s = \frac{(\rho_{sp} - \rho_a)gD_p^2c_2}{18\mu}S_{CF} \qquad (5\text{-}67)$$

where

ρ_{sp} = solid particle density (kg m^{-3})

ρ_a = air density (kg m^{-3})

g = acceleration of gravity (9.81 m s^{-2})

D_p = particle diameter (μm)

μ = 1.81×10^{-5} kg m^{-1} s^{-1} = air viscosity

$$c_2 \quad = 1 \times 10^{-12} \, \text{m}^2 \, \mu\text{m}^{-2} = \text{conversion factor}$$

S_{CF} is an empirical factor:

$$S_{CF} = 1 + 0.16/D_p \tag{5-68}$$

where the units of D_p must be μm in this expression. S_{CF} is between 1.00 and 1.01 for particles with diameters, D_p, greater than 15 μm, and is about 1.16 for particles with diameters of 1 μm.

A plume of large particles with a given diameter, D_p, will "fall away" from the gaseous part of the plume with a speed of v_s, as illustrated in Figure 5-13. The vertical distance "fallen" by the particle plume relative to the remainder of the gas plume at downwind distance, x, is given by $v_s x/u$. This is called the "tilted plume" model in the literature (Hanna et al., 1992). The tilted plume is assumed to have the same shape as the remainder of the gas plume, but is displaced downward due to the settling of the particles.

The deposition of large particles onto the surface can be modeled by assuming that the large particles can be divided into a few size classes, where each size class is characterized by a diameter, D_p, and a concentration, $C(D_p)$, with units mass per unit volume. The local particle deposition flux to the surface, F_{D_p} (with units mass per unit area per unit time) is given by

$$F_{D_p} = v_s C(D_p, x, y, 0) \tag{5-69}$$

where $C(D_p, x, y, 0)$ refers to the ground level concentration of that size class of particles at position (x, y) of the tilted plume. The concentration, C, must be calculated using a dispersion model.

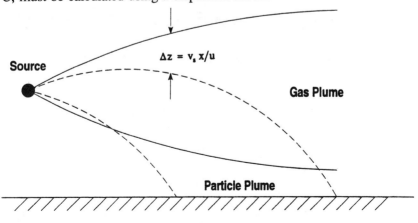

Figure 5-13. Illustration of how a large particle plume will fall away from the rest of the gas plume.

5.9.2. Dry Deposition of Small Particles and Gases

Aside from the large particles, the remaining components of the plume do not have appreciable gravitational settling velocities. However, the plume material can still be removed at the surface by dry deposition, which is caused by chemical attachments to the surface, absorption by vegetation, or other surface processes. Because the distances of interest are usually within a few hundred meters of the source, it is usually assumed that the mass removal due to dry deposition is insignificant when compared to the total mass flux in the plume. A useful procedure for calculating dry deposition is given in the EPA's (1995) revised ISC3 model, which uses the resistance analog, where the deposition velocity is assumed to be inversely proportional to the sum of a set of resistances, as expressed below:

$$v_d = \frac{1}{r_a + r_s + r_t} + v_s \qquad (5\text{-}70)$$

v_s is the gravitational settling speed, which is nonzero for particles (see equations 5-67 and 5-68) and is zero for gases.

The resistances have the following definitions:

r_a = aerodynamic resistance (s m^{-1})

r_s = surface or laminar layer resistance (s m^{-1})

r_t = transfer resistance dependent on surface characteristics (s m^{-1}).

The aerodynamic resistance term is the same for both gases and small particles:

$$r_a = \frac{1}{0.4u_*}\left[\ln\frac{z_d}{z_0} - \psi_H\left(\frac{z_d}{L}\right)\right] \qquad (5\text{-}71)$$

where u_* is friction velocity, L is Monin-Obukhov length, z_d is reference height (assumed to equal 10 m) and the function ψ_H is given by:

$$\psi_H\left(\frac{z}{L}\right) = -5z/L \qquad\qquad 0 < z/L \qquad (5\text{-}72a)$$

$$\psi_H\left(\frac{z}{L}\right) = 0 \qquad\qquad\qquad z/L = 0 \qquad (5\text{-}72b)$$

$$\psi_H\left(\frac{z}{L}\right) = 2\ln\left(\frac{1 +(1 - 16z/L)^{\frac{1}{2}}}{2}\right) \quad z/L < 0 \qquad (5\text{-}72c)$$

Most hazardous gas models will automatically calculate values of u_* and L based on observations of wind speed and stability (see Section 3.3).

The second resistance term in equation (5-70), the surface or laminar layer resistance, r_s, is dependent on the molecular diffusivity of gases or the Brownian diffusivity of particles, and can be estimated from the formula (EPA, 1995):

$$r_s = \left[N_{Sc}^n + \frac{N_{St}}{(1 + N_{St}^2)} \right]^{-1} u_*^{-1} \tag{5-73}$$

where N_{Sc} is the Schmidt number, N_{St} is the Stokes number (equal to 0.0 for gases and defined by equation 5-77 for particles), and

$$n = -0.5 \qquad \text{for } z_0 < 0.1 \text{ m} \tag{5-74a}$$

$$n = -0.7 \qquad \text{for } z_0 < 0.1 \text{ m} \tag{5-74b}$$

The Schmidt number is given by:

$$N_{Sc} = v/D_B \text{ for particles} \tag{5.75a}$$

$$N_{Sc} = v/D \quad \text{for gases} \tag{5.75b}$$

where v is the molecular viscosity of air ($v = \mu/\rho = 0.15 \times 10^{-4}$ m^2 s^{-1}), D_B is the Brownian diffusivity of the particles in air, and D is the molecular diffusivity of the pollutant gas in air. For many gases, N_{Sc} is on the order of unity. For particles, the Brownian diffusivity, D_B, is a strong function of particle size, ranging from $D_B \sim 10^{-11}$ m^2 s^{-1} for $D_p \sim 1$ μm to $D_B \sim 0.1 \times 10^{-4}$ m^2 s^{-1} for $D_p \sim 10^{-4}$ μm. The following formula can be used to calculate Brownian diffusion D_B for particles:

$$D_B = 0.81 \times 10^{-13} \frac{T_{air} S_{CF}}{D_p} \tag{5-76}$$

where T_{air} is air temperature in °K and S_{CF} is the empirical factor given in equation (5-68). The units of D_B are m^2 s^{-1} and the units of particle size D_p are μm.

The Stokes number for particles is given by:

$$N_{St} = \left(\frac{v_s}{g} \right) \left(\frac{u_*^2}{v} \right) \tag{5-77}$$

After substituting typical values for N_{Sc} and N_{St} in equation (5-73), it is evident that r_s is an important term only for gases or very small particles (diameters of 10^{-3} μm or less). r_s can be ignored for particles with sizes of about 1 μm or greater.

The third resistance term in equation (5-70), the transfer resistance r_t, has been the subject of extensive research studies (see the discussions in EPA, 1995) and is generally parameterized by the following formulas:
For particles,

$$r_t = r_a \, r_s v_s \qquad\qquad (5\text{-}78a)$$

For gases,

$$r_t = \frac{1}{LAI/r_f + LAI/r_{cut} + 1/r_g} \qquad\qquad (5\text{-}78b)$$

where the term LAI is the leaf area index defined for vegetated surfaces. LAI is the total area of leaves in the column of air over a unit area of ground surface, r_f is the stomate resistance, and r_{cut} is the cuticle resistance for the vegetation. r_g is the resistance to transfer across the nonvegetated ground or water surface. The first two terms are significant only when vegetation is actively growing and the pollutant is sufficiently reactive to be absorbed by the vegetation. The last term, involving r_g, is significant only if the pollutant is reactive with the surface. For nonreactive gases, the surface transfer resistance, r_t, is infinity and the deposition velocity, v_d, is therefore zero (see equation 5-70).

The terms r_f, r_{cut}, and r_g in equation (5-78b) are well-known only for gases involved in acid deposition processes, such as SO_2, NO_2, HNO_3, peroxyacetyl nitrate (PAN), and O_3. For these gases, Pleim et al. (1984) suggest that r_t is on the order of 10^3 s m^{-1}, with variations of plus or minus a factor of three depending on the particular gas. Little is known about these terms for most hazardous gases. In the absence of information on these specialized terms, it is wise to use the default assumption that the deposition velocity equals 0.01 ms^{-1} (Hanna et al., 1982).

5.9.3. Removal of Particles and Gases by Precipitation and Clouds (Wet Deposition)

Particles and gases can be removed from the plume by rain, snow, clouds, or fog by two mechanisms: (1) in-cloud absorption by small cloud or fog water drops and (2) below-cloud impaction or absorption as large precipitation drops or snowflakes fall through a polluted plume. The first mechanism, in-cloud absorption, is important only for reactive gases and particles, since the ambient water drops are assumed to be *not* moving through the pollutant cloud, and therefore the only way the gases or particles can mix with the drops is by means of an absorption process. As Slinn (1984) explains, if the pollutant and the drops are exposed to each other for a long time, the concentration of chemicals such as SO_2 and NO_2 in the liquid reach an equilibrium determined by Henry's law. This mechanism is clearly different from mechanism (2) above, where the liquid drops fall through the pollutant cloud in a relatively short time (a few seconds, at most) and the primary removal occurs via capture of pollutant particles or aerosols by the droplets.

Both in-cloud and below-cloud removal mechanisms are parametrized in models using a removal scale, Λ (sec)$^{-1}$, which is approximately proportional to the precipitation rate, P (mm/hour). The local concentration, C, is assumed to decrease exponentially with time:

$$C(t) = C(0) \, e^{-\Lambda t} \qquad (5\text{-}79)$$

where t is the time the plume has been exposed to the liquid water drops. The resulting flux of material to the ground, F_{wet}, is given by:

$$F_{wet} = \int_0^{z_w} \Lambda C(z) \, dz \qquad (5\text{-}80)$$

where z_w is the depth of the wetted plume layer. Ramsdell et al. (1993) use the following parameterizations for Λ for iodine gas and aerosol compounds as a function of precipitation rate in their RATCHET model:

$$\text{Rain} \qquad \Lambda = 4 \times 10^{-4} P^{3/4} \qquad (5\text{-}81a)$$

$$\text{Snow} \qquad \Lambda = 6 \times 10^{-5} P \qquad (5\text{-}81a)$$

where precipitation rate, P, is in mm hr^{-1}, and the following rates are suggested.

P (mm hr^{-1} liquid equivalent)			
	Light	Moderate	Heavy
Rain	0.1	3	5
Snow	0.03	1.5	3.3

For example, equation (5-81a) gives $\Lambda \approx 10^{-3}$sec^{-1} or about (15 minutes)$^{-1}$ for moderate rain. This means that most of the pollutant would be removed after being subjected to 15 minutes of moderate rain.

It is suggested that a default value of $(1000 \text{ sec})^{-1}$ be used in codes if the precipitation rate is not known. Equations (5-81a) and (5-81b) can be used if the precipitation rate is known. In the future, as experimental data become available, revised Λ values can be prescribed for specific chemicals. Equation (5-80) is used, knowing Λ and $C(z)$, to calculate the wet flux to the ground.

Because travel times to receptors of interest would be on the order of 10 to 100 sec for expected accidental releases, it is possible to neglect the reduction in total mass flux due to wet deposition of chemicals in the plume. Of course, for larger travel times of 1000 sec or more, it may be necessary to account for the loss of mass.

6

Averaging Times, Concentration Fluctuations, and Modeling Uncertainties

Observed concentrations in the atmosphere are known to be variable in time and space, with turbulent variations similar to those observed for winds, water vapor, and other physical parameters. Several methods for accounting for this variability and its influence on vapor cloud behavior are given below. Most vapor cloud models account for this variability mainly through specification of the averaging time, T_a. The sensitivities of model predictions to variations in averaging time are tested in most of the worked examples in the appendices.

6.1. Overview of Physical Considerations Related to Averaging Time

In order to more efficiently estimate the effects of an accidental vapor cloud release, it is important to have an idea of the conditions associated with types of expected impacts. For example, if propane is released, the likely worst-case effect may be an explosion, which could occur when the instantaneous concentration is in the range from about 2% to 9%. Or, if H_2S is released from a ground-level pipeline rupture a short distance from a group of workers, the worst-case effect would be associated with severe lung damage, which would occur when the 5 to 10 second average concentration exceeded a few thousand ppm. Most accidents that must be dealt with by planners and engineers involve smaller releases of hazardous gases and are concerned with impacts at or beyond the plant "fenceline," at downwind distances of about 1 km. In those cases, the worst-case effect may be, say, minor health impacts when concentrations exceed about 100 ppm for an averaging time of a few minutes. Thus, vapor cloud dispersion calculations may be required for a wide range of averaging times, T_a, depending on the source scenario and the expected impacts. Specific values of T_a should be chosen prior to beginning the modeling analysis, based on a preliminary assessment of the source conditions and the likely worst-case impacts (e.g., flammability versus toxicity, or one second averages versus 10 minute averages).

When concentrations are analyzed using the operational models discussed in Section 7, the interpretation of the results is easier if the duration of the vapor cloud release, T_d, is at least as large as the averaging time, T_a. Otherwise, the concentration time series would contain relatively long strings of zero readings and the toxicological relations that are used may no longer be valid.

Up until about 1990, most vapor cloud modelers did not specifically consider questions related to averaging times. The models implicitly included averaging times to the extent that the dispersion coefficients or entrainment parametrizations used in the models were linked with specific field or laboratory experiments, which themselves were characterized by certain averaging times. For example, the passive gas dispersion coefficients (see Section 5.4) in all vapor cloud models are based on the Pasquill-Gifford-Turner coefficients, which use field experiments with averaging times of about 10 minutes (Hanna et al., 1982). Similarly, many of the models use dense gas entrainment coefficients based on field studies such as the Maplin Sands experiment, with averaging times of about one minute. Another problem sometimes arises when laboratory data have been used to develop these coefficients, since often the scaling relations for averaging times for the laboratory data are uncertain.

Most vapor cloud models now account for averaging time, T_a, in a similar manner in their calculations. This is done in most cases by a simple power-law correction of the form:

$$\frac{C(T_{a1})}{C(T_{a2})} = \left(\frac{T_{a1}}{T_{a2}}\right)^{-\frac{1}{5}} \tag{6.1}$$

This formula was suggested by Turner (1994) in his 1969 edition of the document entitled *Workbook of Atmospheric Dispersion Estimates*. A prescribed minimum for T_a of 20 seconds is often used in equation (6-1), so that concentrations predicted by the equation at small T_a do not rise above known concentrations for instantaneous puffs (see the formulas in Table

5-4). The concentration, C, in this power law formula refers to a position on the mean cloud centerline for that averaging time, and *not* to a position at a fixed geographic location. The concentration also refers to an average over a large ensemble of cases with averaging times of T_a, implying that it is possible that higher (or lower) concentrations would be observed. Equation (6-1) is further described in Sections 6.3 and 6.4.

As suggested above, in any application, it is important for the model user to specify not only the averaging time for the prediction, but also whether the prediction is intended for a cloud centerline (peak) value or a fixed receptor (e.g., the location of a hospital or school). It is also useful to know whether the modeler has intentionally built a degree of conservatism into the model, or whether the model predictions are intended to represent ensemble averages over several realizations of the same conditions.

6.2. Overview of Characteristics of Concentration Fluctuations in Plumes

Most model predictions refer to so-called "ensemble average" concentrations, which represent a statistic such as the most likely (mode) concentration, the median (50th percentile) concentration, or the average concentration expected for the prescribed input conditions. For example, if 1 million field experiments could be performed under identical input conditions (i.e., source emissions rate, wind speed, etc.), the ensemble average would represent the average over the ensemble of all 1 million experiments. Because the atmosphere is always turbulent, the observations from any given "realization" or field experiment are likely to be different from the mean, as demonstrated by the typical time series of concentrations shown in Figure 6-1, for which the peak concentration, C_{peak}, for averaging time, T_a, is about 1.8 times the mean, C. This peak value at a given T_a is obviously affected by the sample size (e.g., the peak value out of 10 numbers is likely to be less than the peak value out of 1000 numbers taken from the same basic distribution).

Chatwin (1982) points out that in many cases involving accidental releases of flammable gases such as propane, the maximum short term (~1 sec) concentration is the most important variable to predict. According to Chatwin, the probability density function (pdf) of concentration fluctuations in the atmosphere is usually characterized by a standard deviation, σ_c, at least as large as the mean. The pdf describes the fraction of concentrations occurring for given narrow ranges of concentration values. The relative

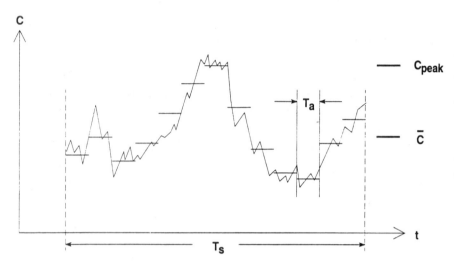

Figure 6-1. Illustration of concentration time series with sampling time, T_s, and averaging time, T_a, indicated. Concentrations averaged over time T_a are shown as short horizontal lines. The peak concentration, C_{peak}, over averaging time, T_a, and the mean, $\overline{C}$, for this time series are shown.

magnitude of concentration fluctuations ($\sigma_c/\overline{C}$) is observed to be the same order as the relative magnitude of velocity fluctuations (σ_u/u) in the atmosphere. The parameters σ_c and σ_u are the standard deviations of turbulent fluctuations in concentration and wind speed, respectively. For the release scenarios involving flammable or toxic gases, it is important to predict the upper end of the pdf (i.e., the highest expected concentrations).

Wilson (1995) states that consideration should also be given to the time scales of the turbulent concentration fluctuations. The question often comes up that, given a current value of the concentration, how long will the concentration exceed a certain threshold during a future time period? Or, how many times will the threshold be exceeded over the duration of the incident? Some parametrizations that answer these questions are given in Section 6.5.

6.3. Predictions of Concentrations on the Plume Centerline at a Given Downwind Distance as a Function of Averaging Time, T_a

Hazardous gas models can predict the crosswind distribution of ensemble mean concentrations at distance x from the source for a certain averaging

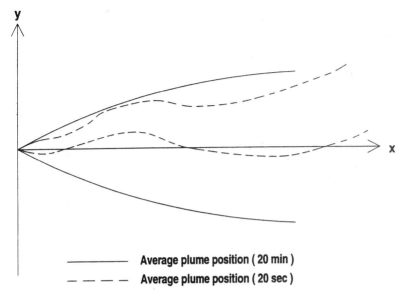

Figure 6-2. Comparison of short- and long-term average plume position.

time, T_a. In most dense gas modules, these fundamental model predictions are appropriate for averaging times of about one or two minutes, which correspond to the characteristics of the field data on which the dense gas model parameters are based. These corresponding model predictions of the ensemble average maximum concentration on the plume centerline, $C_{cl}(x,T_a)$, are not keyed to any particular geographic point—the only restriction is that the downwind distance must be x. But because natural plumes meander or swing back and forth, the ensemble average centerline concentration will decrease as averaging time increases, and the geographic position of the centerline is likely to shift as T_a varies. Figure 6-2 compares the shape of short- and long-term average plumes, showing how the centerline positions vary and suggesting that the maximum concentrations would be higher in the narrower, short-term average plume.

These concepts can be better explained by considering an ensemble or large number of concentration observations under certain initial and boundary conditions. Then the variation of the distribution of the centerline concentrations, C_{cl}, with T_a at a fixed downwind distance, x, would be as shown in Figure 6-3. A box plot format is used, where key points on the probability density function (pdf) at each T_a are indicated. The dashed line on the figure represents the ensemble average model prediction, which passes through the medians of the distributions. If the model predictions are

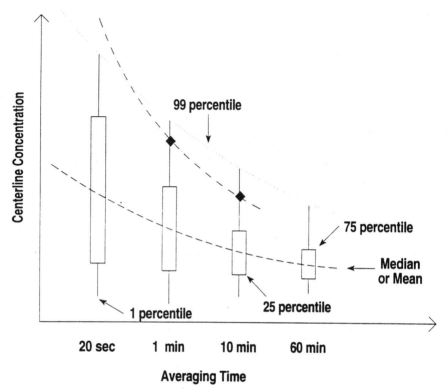

Figure 6-3. Typical distributions of maximum concentration on plume centerline, C_{cl}, at a given x, for various averaging times, T_a. The dashed line goes through the means at each T_a, and the dotted line goes through the 99th percentile of each distribution. The dashed–dotted line goes through the percentile associated with the maximum at that T_a assuming the sampling time is 60 min, and is given by the formula $100[1 - (T_a/60 \text{ min})]$.

corrected for averaging time, T_a, the corrected ensemble average concentrations should fall along this dashed line. As averaging time, T_a, approaches 0.0 (i.e. an instantaneous snapshot of the plume), the concentration C_{cl} should approach a value representative of the instantaneous plume.

It should be mentioned that some models may be designed to be slightly conservative—that is, to predict plume centerline concentrations, C_{cl}, higher than the median. For example, if a model were designed to predict the 99th percentile of the concentration distribution at each T_a, the concentration predictions would follow the dotted line in Figure 6-3. If a model were designed to give the maximum at a given T_a for a given total sampling time, T_s, (e.g., for the total time record of data being analyzed) which is 60 min for the example in the figure, the concentration predictions would follow the dash–dot line. In this latter example, where the total number of concentration values equals 60 min/T_a, the percentile associated with the single maximum concentration would increase as T_a decreases. Note that the dash–dot curve is discontinued for T_a exceeding about 20 min, since there are only three 20 min and only two 30 min averages that can be calculated from a 60 min time series.

Most hazardous gas models that correct for averaging time are attempting to follow the median (dashed line in Figure 6-3), even though they may not articulate these conditions. In addition, as mentioned at the end of Section 6.1, most models accomplish this correction to the concentration by applying a $T_a^{1/5}$ power law correction to the lateral dispersion coefficient, σ_y:

$$\frac{\sigma_y (T_{a2})}{\sigma_y(T_{a1})} = \left(\frac{T_{a2}}{T_{a1}}\right)^{1/5} \tag{6-2}$$

Then assuming that cloud centerline concentration is inversely proportional to σ_y, equation (6-1) follows from equation (6-2). In order to prevent σ_y from dropping below its observed value (Slade, 1968, see Table 5.4) for instantaneous conditions, which would inevitably happen with equation (6-2) as $T_{a2} \to 0$, a "minimum T_{a2}" criterion is usually applied. This is the T_{a2} which would result in σ_y equaling the following values given by Slade (1968) for instantaneous plumes or puffs:

$$\text{Unstable} \qquad \sigma_{yI} = 0.14x^{0.92} \tag{6-3}$$

$$\text{Neutral} \qquad \sigma_{yI} = 0.06x^{0.92} \tag{6-4}$$

$$\text{Very Stable} \qquad \sigma_{yI} = 0.02x^{0.89} \tag{6-5}$$

For neutral conditions, this criterion is satisfied at T_{a2} equal to about 20 seconds, where it is assumed that σ_y for continuous plumes is given by the standard EPA formulas given in Table 5-1. However this minimum T_{a2} is dependent on what is assumed for (1) distance x, and (2) representative averaging time for the formulas in Table 5-1. Furthermore, equations (6-3)–(6-5) themselves are based on limited data and would have significant uncertainties (say ±50%).

As a default parametrization, a procedure that accounts for averaging time is proposed based on the following assumptions:

- The EPA formulas for σ_y for continuous plumes are valid for an averaging time of 10 minutes.
- The "minimum T_a" criterion is 20 seconds.
- Equation (6-2) is valid for σ_y corrections for T_a.

Some models such as DEGADIS and HEGADAS assume that the lateral distribution of concentration in a dense gas plume is made up of a dense gas core (with constant concentration) of width W, and Gaussian edges with standard deviation, σ_y. The averaging time correction is then applied by these models only to the Gaussian edges. Hanna et al. (1993) showed that this approach led to large underestimates of observed variations of dense gas concentrations with averaging time. Consequently, the approach described by the three bullets above assumes that the averaging time corrections apply to the entire plume width, as written in equation (6-2).

It can be assumed, from the Gaussian plume formula, that concentration C is inversely proportional to $\sigma_y \cdot \sigma_z$, and that σ_z is unaffected by averaging time. This latter assumption is justified by the fact that vertical dispersion is caused by turbulent fluctuations that have much smaller time scales than lateral dispersion, and thus the vertical dispersion coefficient, σ_z, is not influenced by the larger meandering turbulent eddies that cause σ_y to vary with averaging time. The general power law given in equation (6-1) is therefore justified for concentration estimation.

Equation (6-1) applies to the mean or median (i.e., the dashed line) plume centerline concentration in Figure 6-2. If there is interest in the centerline concentration at a given averaging time at any percentile on the distribution function (e.g., see the dotted line representing the 99th percentile on Figure 6-2), an assumption is needed for the form of the pdf. For in-plume fluctuations with intermittency $I = 1$ (i.e., the concentration is always greater than zero), a log-normal distribution can be used (see Hanna, 1984, and Wilson, 1995):

$$p(\ln C') = \frac{1}{\sqrt{2\pi}\sigma_{\ln C'}} \exp\left[\frac{-(\ln C' - \overline{\ln C'})^2}{2\sigma_{\ln C'}^2}\right] \tag{6-6}$$

$$P(\ln C) = \int_{-\infty}^{\ln C} p(\ln C')\, d(\ln C') \tag{6-7}$$

where p is the probability density function and P is the cumulative density function (ranges from 0.0 to 1.0).

At small averaging times ($T_a \sim 20$ seconds or less), atmospheric data presented by Hanna (1984) show that

$$\frac{\sigma_{\ln C}}{|\overline{\ln C}|} \approx 1.0 \tag{6-8}$$

It can be assumed that $\sigma_{\ln C}$ decreases as averaging time increases according to the following approximation:

$$\frac{\sigma_{\ln C}^2 (T_a)}{\sigma_{\ln C}^2 (20 \text{ sec})} = \frac{1 + 20 \text{ sec}/2T_I}{1 + T_a/2T_I} \qquad (T_a > 20 \text{ sec}) \qquad (6\text{-}9)$$

where T_I is the integral scale for turbulent fluctuations in concentration. For plumes in the atmospheric boundary layer, a default assumption would be

$$T_I \approx 300 \text{ seconds (Default)} \qquad (6\text{-}10)$$

With this value of T_I, equation (6-9) gives

$$\sigma_{\ln C} \text{ (one hour)} = 0.4\sigma_{\ln C} \text{ (20 sec)} \qquad (6\text{-}11)$$

These formulas can be used to estimate the variations of mean centerline concentration with averaging time and to determine the probability of certain high concentrations occurring for that averaging time.

6.4. Predictions of Concentrations at a Given Receptor Position as a Function of Averaging Time, T_a

The discussions in the previous subsection were concerned with predicted concentrations on the plume centerline or axis, which can shift position with time. For that type of model application, the analyst is concerned only with the maximum plume impact at a given downwind distance independent of location. Another type of model application would be concerned with the plume impact at a given receptor position, as defined by for example a monitoring site or a critical subset of the surrounding population (say a school or a hospital).

Consider an ensemble of concentration observations from a given monitoring site. The data can be assumed to be taken from many independent field studies, all with nearly the same ambient conditions (i.e., release rate, wind speed and direction, stability). These observations would show a variation of probability density functions (pdf's) with averaging time as suggested in Figure 6-4. Note that there are three major differences between Figures 6-3 and 6-4:

Figure 6-3 C on Plume Centerline	Figure 6-4 C at Fixed Receptor
Median C decreases as T_a increases	Median C is constant with T_a
There are no zeros in C time series	There are many zeros in C time series
σ_C is relatively small	σ_C is relatively large

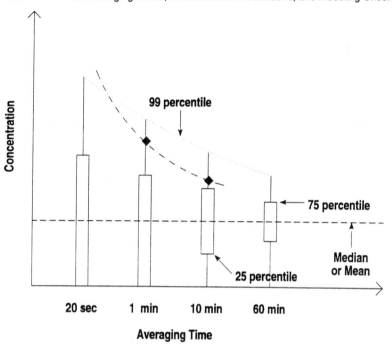

Figure 6-4. Typical distributions of concentration observed at a given monitor or receptor location, for various averaging times, T_a. The dashed line goes through the means at each T_a, and the dotted line goes through the 99th percentile of each distribution. The dashed-dotted line goes through the maximum at that T_a assuming the sampling time is 60 min, and is given by the percentile $100[1 - (T_a/60 \text{ min})]$.

All of these differences are due to the fact that, in the case of Figure 6-4, the plume can meander away from the receptor, leading to the possibility of $C = 0$ observations during part of the time period at that receptor. In contrast, by definition C_{cl} is always greater than zero in Figure 6-3. It is noted that most dense gas models are intended to apply to the scenario whose concentration pdf's are plotted in Figure 6-3, where the primary interest concerns the concentration on the plume centerline.

Often the variation of C_{max} with T_a is calculated from data at fixed receptors. A time series $C(t)$ is searched in order to identify the various $C_{max}(T_a)$; for example, this was done by Hanna et al (1993) using the field data from the Burro, Coyote, and Desert Tortoise experiments. The resulting C_{max} values would follow the dot–dash curve in Figure 6-4. In the example used to generate this curve, the total length or sampling times, T_s, of the time series is assumed to be 60 min. If the data were divided into consecutive blocks in which the concentration were averaged over T_a (min), there would be 60 min/T_a numbers available at each T_a. The percentile associated with the C_{max} for each T_a is then given by:

$$\frac{\text{Percentile}}{100} = 1 - \frac{T_a}{60 \text{ min}} \tag{6-12}$$

Thus the maximum concentration, C_{max}, for $T_a = 1$ min would be associated with a percentile of $(1 - 1/60)$, or 98.3%.

As T_a exceeds 10 min and approaches 60 min, this formula and concept breaks down, since there are so few numbers involved (e.g., there are only two 30 min average concentrations that can be calculated from a 60 min time series). Note on Figure 6-4 that the slope of the curve showing the variation of C_{max} with T_a is greater than the slope of the curve showing the variation of C (fixed percentile) with T_a. From a theoretical point of view, C (fixed percentile) is preferable, but from a practical point of view researchers usually work with C_{max}. It is clearly important to at least recognize the difference.

As suggested by Wilson (1995), there is sometimes confusion generated by authors who publish so-called $C_{max}/\overline{C}$ data and who use these data to estimate the exponent in the power law in equation (6-1). This power law should be valid for the ensemble-mean centerline concentration (the dashed line in Figure 6-3); however, many persons mistakenly calculate the slope of the dot–dash curve in the figure (i.e., the peak observed concentration for a given averaging time). Wilson reanalyzes several older data sets in order to demonstrate the implications of these misinterpretations.

The distribution function for the fixed receptor data in Figure 6-4 must account for the possibility of many zeros, since the instantaneous plume may sometimes meander away from the receptors. Wilson (1995) suggests several alternative pdf formulas, including the exponential and the clipped normal. The formulas valid for the exponential distribution are given below:

$$p(C) = \frac{I^2}{\overline{C}} \exp\left(-\frac{IC}{\overline{C}}\right) + (1 - I)\delta(0) \tag{6-13}$$

$$P(C) = 1 - I \exp\left(\frac{-IC}{\overline{C}}\right) \tag{6-14}$$

where the cumulative density function P in equation (6-14) is obtained by integrating the probability density function p in equation (6-13) from 0 to C.

$$\frac{\sigma_c}{\overline{C}} = \left(\frac{2}{I} - 1\right)^{1/2} \tag{6-15}$$

where δ is the Dirac delta function (equal to 1.0 at $C = 0.0$ and equal to 0.0 at $C > 0$), and I is the so-called intermittency, or fraction of nonzero observations in the total record ($I = 1.0$ if the plume is always impacting the receptor). A typical value of I in the atmosphere for sampling times of about 30 minutes and averaging times of 10 to 20 seconds is about 0.2. In the absence of other information, a default value of $I = 0.2$ is recommended for very small averaging times, T_a, leading to the formulas:

$$\left.\begin{array}{c} P(C) = 1 - 0.2 \exp\left(-0.2\dfrac{C}{\overline{C}}\right) \\[2em] \dfrac{\sigma_c}{\overline{C}} = \left(\dfrac{2}{I} - 1\right)^{1/2} = 3 \end{array}\right\} \text{ as } T_a \to 0 \qquad (6\text{-}16)$$

As averaging time increases to 60 minutes, equation (6-9) can be used to calculate $\sigma_c^2(T_a)/\sigma_c^2$ (20 sec), again assuming that the integral time scale, T_I, is 300 seconds. The intermittency, I, can be calculated by inverting equation (6-15):

$$I = \frac{2}{1 + (\sigma_c/\overline{C})^2} \qquad (6\text{-}17)$$

Specifically, the procedures are given below to estimate the probability of certain high concentrations occurring at a fixed receptor for a given averaging time.

Step 1: Calculate

$$\frac{\sigma_c^2(T_a)}{\sigma_c^2 \,(20 \text{ sec})} = \frac{1 + 20 \text{ sec}/600 \text{ sec}}{1 + T_a/600 \text{ sec}}$$

Step 2: Calculate

$$I(T_a) = \frac{2}{1 + (\sigma_c/\overline{C})^2}$$

Step 3: Calculate

$$P(C) = 1 - I \exp\left(\frac{-IC}{\overline{C}}\right)$$

It is assumed that $\overline{C}$ is known. Equations (6-14) through (6-17) should not be used at $T_a > 3600$ sec, since the intermittency, I, would be calculated to exceed 1.0, which is impossible. Instead, use $I = 1.0$ and $\sigma_c/\overline{C} = 1.0$ at $T_a > 3600$ sec.

These formulas can be used to estimate the probability of certain high concentrations occurring at a given receptor position for a given averaging time.

6.5.   Threshold Crossing Probability

The probability density functions (pdf's) discussed in Sections 6.3 and 6.4 do not contain any information on the expected *time interval* between crossings of certain concentration thresholds. For example, if the observed HF concentration is dropping below 100 ppm now, what is the time period before it exceeds 100 ppm again? None of the vapor cloud models include this capability, although this topic is the subject of some current ongoing research (see the review by Wilson, 1995). The problem can best be seen by studying a concentration time series such as that shown by Wilson (1995) in Figure 6-5. For this example, the concentration moves up across the threshold concentration, C_*, three times. The return time, t_R, is defined as the time between two successive C_* up-crossings. After C_* is crossed, the concentration remains above C_* for a time, t_{above}. Clearly the toxicological effect will be larger as the time, t_{above}, increases.

Wilson (1995) provides discussions and tentative formulas for many statistics related to threshold crossings. Because these formulas are complicated and often depend on poorly known parameters such as the integral time or distance scale of the concentration fluctuations, they are not given here. The reference, which is available from the AIChE/CCPS, provides extensive details for interested readers.

6.6.   A General Structure for the Analysis of Model Uncertainties

If major decisions are to be made regarding mitigation procedures, evacuation plans, and risk assessments regarding hazardous gases, it is important to have the best possible confidence in the models that are used and the data that are being collected. It may even be possible to build the confidence intervals (uncertainty) into the decision-making process. The purpose of this subsection is to discuss the three components of total error or uncertainty in models for source emissions, transport and dispersion of hazardous gases:

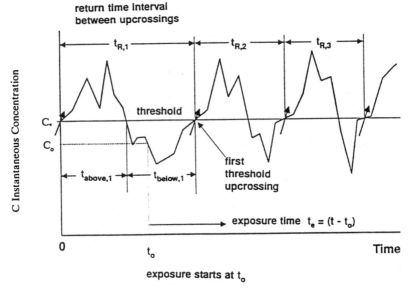

Figure 6-5. Schematic diagram illustrating variation in return time between upward crossings of threshold C. (from Wilson, 1995).

I. Errors caused by model physics assumptions

II. Random variability (e.g., concentration fluctuations)

III. Uncertainties generated by input data errors

These components have not yet been studied in any comprehensive way. A general philosophy on model evaluation and uncertainties is shown in Figure 6-6, where the three components of uncertainty defined above are plotted as a function of the number of parameters in the model. The total uncertainty is shown to first decrease and then increase as the number of parameters rises, with the increase primarily due to errors in input data. It is desirable to construct a model such that the total model uncertainty on the figure is minimized. The uncertainties can be quantified by defining total uncertainty as $\overline{(C_p - C_O)^2}$ and assuming that C_p and C_O are given by:

$$C_O = \overline{C_O} + C'_O + \Delta C_O \tag{6-18}$$

$$C_p = \overline{C_p} + C'_p + \Delta C_p \tag{6-19}$$

where

$\overline{C_O}$ is the actual ensemble average

C'_O is the stochastic (random) variability

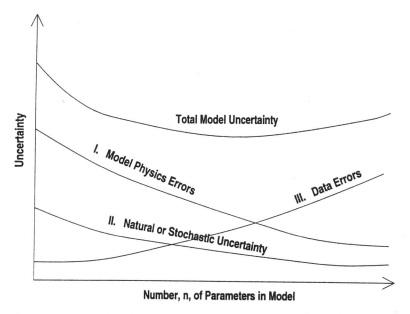

Figure 6-6. Illustration of variation of model uncertainty components with number of parameters in model.

ΔC_O is the data error in C_O
$\overline{C_p}$ is the predicted ensemble average
C'_p is the predicted random variability
ΔC_p is the error due to data input errors

Except for SCIPUFF (Sykes et al., 1993), none of the hazardous gas models that are in common use (see Section 7) are able to predict the random component C'_p. If it is assumed that there is no correlation among the components, then equations (6-18) and (6-19) can be used to calculate the total uncertainty, $\overline{(C_p - C_O)^2}$ which can be expressed by the sum of the three major components discussed above.

$$\overline{(C_p - C_O)^2} = \overline{(\overline{C_p} - \overline{C_O})^2} + \sigma_{C_O}^2 + \overline{\Delta C_O^2} + \overline{\Delta C_p^2} \qquad (6\text{-}20)$$

Total Model	*Model Physics*	*Stochastic*	*Data*
Uncertainty	*Error*	*Uncertainty*	*Errors*
I	II	III	

As shown in subsections 6-3 and 6-4, the stochastic uncertainty component II is roughly equal to the square of the mean (i.e., $\sigma_{C_O}^2/\overline{C_O^2} \approx 1$) and can be predicted by, for example, equations (6-9), (6-14), and (6-15). The data errors component III is also believed to be of the same order as the square of the mean. The model physics error component I can be estimated from the results of the model evaluation exercises such as those discussed in Section 8, which give the total model uncertainty. Then component I is calculated through equation (6-20), assuming components II and III have been determined.

The data errors component III can be estimated based on studies of instrument errors in the field. It is stressed that some QC (quality control) procedures, such as checking the voltage output of an anemometer at a given rotation rate, do not tell much about actual instrument error *in the field.* These actual errors are best determined through use of co-located instruments and comparison with high quality "baseline" instruments. The $\overline{\Delta C_0^2}$ term in equation (6-20) is simply the error in the concentration measurement. If there is a minimum error of 10% in the observation of C, as is typical for many instruments, then the total model error in a model evaluation exercise clearly has a minimum of 10%. Because the model prediction C_p can always be expressed as $C_p = f(x_1, x_2, x_3, \ldots, x_n)$, the term $\overline{\Delta C_p^2}$ in equation (6-20) can be calculated from:

$$\left(\frac{dC_p}{C_p}\right)^2 = \left(\frac{\partial f}{\partial x_1} \frac{dx_1}{x_1}\right)^2 + \left(\frac{\partial f}{\partial x_2} \frac{dx_2}{x_2}\right)^2 + \cdots \qquad (6\text{-}21)$$

where it is assumed that the dx_i terms are not correlated.

As an example of the use of this method, assume a simple formula for pollutant concentration:

$$C_p = \frac{AQ}{u} \qquad (6\text{-}22)$$

where A is a constant, Q is source emission rate, and u is wind speed. Then equation (6-20) yields:

$$\overline{(dC_p/C_p)^2} = \overline{(dQ/Q)^2} + \overline{(du/u)^2} \qquad (6\text{-}23)$$

If the observed data errors are assumed to be 3% for (dQ/Q) and 10% for (du/u), then the expected error in the predicted concentration is $[(3\%)^2 + (10\%)^2]^{1/2}$, or 10.4%. The reader can quickly see that the total error for a model with many data inputs can grow to be quite large.

A more computer-intensive way of assessing the model uncertainties due to sensitivities to variations in input parameters is to run the model thousands of times, each time randomly selecting input parameters from their known range of uncertainty, but preserving correlations among the input parameters. This is known as the Monte Carlo method because of the random nature of the data selection procedures.

It is important to recognize that no matter how good the hazardous gas model is, it is still subject to the effects of data input uncertainties and stochastic variabilities. The model evaluation exercise described in Section 8 demonstrates that, once models achieve a certain accuracy (say, mean biases of ±20% and scatter of 50%), they cannot be improved any further due to these fundamental uncertainties.

7

Overview of Operational Vapor Cloud Models in Common Use

7.1. Summary of Commonly Used Models

As mentioned earlier, there are hundreds of accidental release and dispersion models available, for application to dense gas releases. However, only about 20 to 25 of these are in wide use at the time of this writing (April, 1995), as seen from the results of the model survey reported in Section 7.2. The characteristics of these models can be compared in two primary ways: (1) planning versus real-time models, or (2) public versus proprietary models.

All 22 models reviewed in this section can be used in the planning mode, where the software is run by trained engineers for a variety of trial scenarios. The results are then used to make decisions regarding mitigating measures such as reduction in inventory, installation of emergency block valves, construction of different-sized storage tanks, modifications of specifications for tanks, pipelines, or dikes, and installation of water sprays or vapor barriers. Because these planning systems consist solely of computer software and do not require additional hardware, their costs generally do not exceed $10,000 or $20,000.

In contrast to the planning model systems, real-time or emergency-response modeling systems consist of a combination of software and hardware. The software would be installed on a dedicated computer, located on-site and operated continuously. Updated data such as storage tank conditions and local meteorology would be available to the system such that when an accidental release might occur, the operators would spring into action and immediately run the model using the latest inputs. The resulting predictions of vapor cloud movement and concentrations would be used to make decisions regarding emergency response. Only a few of the models reviewed in this section can be applied as real-time systems (e.g., CHARM, CHEM-MIDAS, SAFER, ALOHA, or HOTMAC/RAPTAD). Except for the U.S. government-developed CAMEO/ALOHA model, the costs for these total systems (including hardware) can be well over $100,000. Over the past five years, there has been a deemphasis on these real-time systems, partly because they are difficult to use in the confusion of an accident and are sensitive to errors in input data, and partly because many chemical industries have decided to shift their emphasis to "planning" studies.

Another way to divide the models is according to whether they are public or proprietary. The public models' source codes and technical documentation are generally available at a cost that does not exceed the actual cost required to copy and distribute the information (usually less than $1000). On the other hand, because the proprietary models were developed by private companies in order to generate income, their source codes are not available, the technical documentation is often not complete, and the costs are determined by market competition (costs are usually in the range from $10,000 to $20,000). The 22 models in Section 7.2 can be divided by this criterion in the following way:

Public: AQPAC, ALOHA, DEGADIS, DRIFT, FEM3C, HGSYSTEM, PLM89A, SCIPUFF, SLAB, TSCREEN, VDI Guideline

Proprietary: AIRTOX, CANARY, CHARM, CHEM-MIDAS, GASTAR, HOTMAC/RAPTAD, PHAST, SADENZ/SACRUNCH/SAPLUME, SAFEMODE, SuperChems Expert, TRACE

It is emphasized that these 22 models are the ones for which question-naires were completed; a few other models such as ADAM (Raj and Morris, 1988; Raj, 1990), could have been included, but no current information was sent to us on those models. Nevertheless, we feel that the 22 models comprise a representative set of available models and do include all widely used models as of April 1995.

References for the publicly available models are listed below. Documentation also exists for the proprietary models, but it is available only to purchasers.

AQPAC   (Daggupaty, 1988)
ALOHA   (NOAA, 1992; NOAA/HMRAD and EPA/CEPPO, 1992)
DEGADIS (Havens and Spicer, 1985; Havens, 1988; Spicer and Havens, 1990)
DRIFT   (Webber et al., 1992; Jones et al., 1993)

FEM3C	(Chan, 1994)
HGSYSTEM	(Witlox, 1994a and 1994b; Witlox and McFarlane, 1994; Post, 1995)
PLM89A	(Bloom et al., 1989)
SCIPUFF	(Sykes et al., 1993)
SLAB	(Ermak, 1990)
TSCREEN	(EPA, 1994)
VDI Guideline	(VDI, 1990)

On average, the proprietary models are found to be more user-friendly and to have a larger chemical data base. They also are generally more comprehensive in that they all handle *both* source emissions and transport and dispersion calculations. On the other hand, the proprietary models tend to be "black boxes" that do not encourage research-type studies. In contrast, some of the public models are not so user-friendly, and many of them do not calculate source emissions at all. Despite this limitation, many of the public models contain state-of-the-art scientific algorithms that are very useful for carrying out detailed technical analyses that can be easily communicated to other scientists and engineers in industry and in government agencies. Because the regulatory agencies developed some of these models (e.g., the EPA developed TSCREEN), the results of calculations with these models are likely to be better recognized by the agencies.

Users of vapor cloud models are cautioned that, since there are no regulatory guidelines for these types of models, their input requirements and output conditions may not be consistent. The user may need to do additional analysis or write pre- or post-processing software in order to obtain the desired output information from the available input information. In particular, source emissions (i.e., mass release rate) must be calculated by the user if models such as SLAB and DEGADIS are used, which have no source emissions algorithms (as illustrated in all of the worked examples in Appendices A through G). The users must apply the formulas in Chapter 4. Or, on the output end of the model, the user may have to carry out independent analyses in order to generate, say, contour plots or vertical profiles.

7.2. Characteristics of Commonly Used Vapor Cloud Dispersion Models

A questionnaire on vapor cloud dispersion models was sent to over 100 developers of such models. In order to reduce the total population of models being considered, the restriction was imposed that the model must be applicable for dense gas releases (this restriction eliminates models such as ISC, AFTOX, and Shell-SPILLS). Responses were received on 22 models. Table 7-1 contains a listing of the names of these 22 models and the addresses of the contact persons. All of these models are available as software and are backed up by user's guides or technical documentation. Table 7-2 contains the modelers' responses to the questionnaire, indicating the capabilities of each model. Note that these responses have not been edited by us, and the information pertains to model versions available in April 1995. Readers are urged to contact the developers if they would like more details.

The models in Table 7-1 all contain algorithms for calculating atmospheric transport and dispersion. In addition, questionnaires were completed by several groups for models that are restricted to source term calculations (e.g., evaporation rate from a liquid sill, or mass flow rate from a pressurized tank rupture). Discussions of the characteristics of these source term models have been given in Chapter 4.

The reader should realize that many models are currently being updated. The models described in Tables 7-1 and 7-2 represent the versions available in April 1995. The developers should be contacted to obtain the most recent version of their models.

TABLE 7-1
Listing of Vapor Cloud Models and Addresses of Developers or Contact Persons
These are the Models for which Questionnaires were Completed
(See Table 7-2 for the Results from the Questionnaires)

Model Name		Public or Proprietary
AIRTOX	M. Mills, ENSR, 35 Nagog Park, Acton, MA 01720	Proprietary
ALOHA	J. Galt, HMRAD/NOAA, 7600 Sand Point Way NE, Seattle, WA 98115	Public
AQPAC	S. Daggupaty, AES, 4905 Dufferin St., Downsview, Ontario, Canada M3H 5T4	Public
CANARY	K. Cromwell, Quest, 908 26th Ave. NW, Norman, OK 73070-8069	Proprietary
CHARM	M. Eltgroth, Radian, POB 201088, Austin, TX 78720-8069	Proprietary
CHEM-MIDAS	K. Woodward, PLG, 7315 Wisconsin Ave., Ste 620 East, Bethesda, MD 20814-3209	Proprietary
DEGADIS*	T. Spicer, Un. Ark., CHRC, Fayetteville, AR 72701	Public
DRIFT	S. Jones, AEA Tech, Warrington, Cheshire, UK WA3 6AT	Public
FEM3C	S. Chan, L-262, POB 808, LLNL, Livermore, CA 94551	Public
GASTAR	J. Handley, CERC, 3 King's Parade, Cambridge, UK CB2 1SJ	Proprietary
HGSYSTEM*	P. Roberts, Shell Labs, P.O.B. 1, Chester, UK CH1 3SH	Public
HOTMAC/RAPTAD	T. Yamada, YSA, Rt. 4, Box 81-A, Santa Fe, NM 87501	Proprietary
PHAST	M. Johnson, DNV Technica, 40925 County Center Dr., Ste 200, Temecula, CA 92591	Proprietary
PLM89A	W. Goode, LMES, POB 2003, Oak Ridge, TN 37831-7212	Public
SADENZ/SACRUNCH/ SAPLUME	G. Kaiser, SAIC, 11251 Roger Bacon Dr., Reston, VA 22090	Proprietary
SAFEMODE	P. Raj, TMS, 99 S. Bedford St. Ste 211, Burlington, MA 01803	Proprietary
SCIPUFF	J. Hodge, DNA, 6801 Telegraph Rd., Alexandria, VA 22310-3398	Public
SLAB*	D. Ermak, Lawrence Livermore Nat. Lab., P.O.B. 808, Livermore, CA 94550	Public
SUPER CHEMS EXPERT	G. Melhem, ADL, 20 Acorn Park, Cambridge, MA 02140	Proprietary
TRACE	S. Khajehnajafi, SAFER, 4165 E. Thousand Oaks Blvd, Ste 350, Westlake Village, CA 91362	Proprietary
TSCREEN*	J. Touma, USEPA, MD-14, Research Triangle Park, N.C. 27711	Public
VDI	M. Schatzmann, Un. Hamburg, Bundastrasse 55-D, 20146 Hamburg, Germany	Public

*Can be obtained from EPA TTN Bulletin Board (Phone No. 919-541-5742)

TABLE 7-2
Results from Model Questionnaires (Most Answers Are Given as Y = Yes or N = No)

Question	AIRTOX	ALOHA	AQPAC	CANARY	CHARM	CHEM-MIDAS	DEGADIS	DRIFT	FEM3C	GASTAR	HGSYSTEM
Operating Information											
Main use: R=Research; A=Applied	A	A	A	R,A	A	R,A	R,A	A	R	A	R,A
Number sold or given away?	—	2500	80	—	250	12	—	—	—	—	100
Input Data											
Accept real time weather data?	N	Y	Y	N	Y	N	N	N	N	N	N
Method of data entry: H=Hand; F=data file memory; D=disk or tape	H,F,D	H	H,F	H,F,D	H,F,D	H,F	H,F	H,F	F	H,F	H,F
Number of chemicals hardwired (I=input by user)	94	492	70	300	150	80	7	25	12	—	30
Source Emissions Models?											
Evaporation of Spilled Liquids?	Y	Y	Y	Y	Y	Y	N	N	Y	Y	Y
Flashing	Y	Y	Y	Y	Y	Y			N	Y	Y
Multicomponents	Y	Y	N	Y	Y	Y			Y	Y	Y
Entrainment as aerosols?	N	N	N	Y	Y	N			Y	Y	Y
Heat transfer, substrate to cloud?	Y	Y	N	Y	Y	Y			Y	Y	Y
Mass transfer in liquid phase?	Y	Y	Y	Y	Y	Y			N	Y	Y
Gas flux from container rupture?	Y	N	Y	Y	Y	N			N	Y	Y
Transport and Dispersion Model?	Y	Y	Y	Y	Y	Y	Y	Y	Y	Y	Y
Releases treated: I=Instantaneous; C=Continuous; V=Variable	I,C,V	I,C,V	I,C	I,C,V	I,C,V	I,C,V	I,C,V	I,C,V	I,C,V	I,C,V	I,C,V
Dense, Neutral, or Buoyant?	D,N,B	D,N	D,N,B	D,N,B	D,N,B	D,N,B	D,N,B	D,N	D	D,N,B	D,N,B
Jet	Y	N	N	Y	Y	Y	Y	N	N	Y	Y
Evaporation of aerosols	Y	Y	N	Y	Y	Y	Y	Y	N	Y	Y
Condens. of moisture in vapor cloud?	Y	Y	N	Y	Y	Y	Y	Y	Y	Y	Y
Surface roughness?	Y	Y	Y	N	Y	Y	Y	Y	Y	Y	Y
Complex terrain handled?	N	N	N	N	Y	N	N	N	Y	N	Y
U variations in time and space?	time	N	N	N	Y	Y	N	N	N	N	N
Indoor concentrations?	Y	N	Y	N	N	N	N	Y	Y	N	N
Building wake effects?	N	Y	N	N	N	Y	N	N	N	N	N
Dispersion Model: B=Box or Slab; G=Gaussian; K=K/numerical	B,G	G,K	B,G,K	B,G	Puff	G,B	K	B	K	B	B,G
Along-wind dispersion?	Y	Y	Y	Y	Y	N	Y	Y	Y	Y	Y
Lift-off?	Y	N	N	Y	Y	N	N	N	Y	N	Y
Chemical reactions in plume?	N	N	N	N	Y	N	Y	Y	Y	N	Y
Dry or wet deposition?	N	N	Y	N	N	N	N	N	Y	N	Y
Concentration fluctuations?	N	N	N	N	Y	Y	N	N	N	N	Y
Output											
Averaging time (minutes) (I=input)	10 min	Varies	10 min	—	10 sec–10 min	5 min	—	—	—	—	—

Question	HOTMAC/RAPTAD	PHAST	PLM 89A	SADENZ/ SACRUNCH/ SAPLUME	SAFE-MODE	SCIPUFF	SLAB	SUPER CHEMS EXPERT	TRACE	TSCREEN	VDI
Operating Information											
Main use: R=Research; A=Applied	R,A	A	A	R,A	A	A	A	R,A	R,A	A	A
Number sold or given away?	—	350	—	—	—	—	—	53	—	—	300
Input Data											
Accept real time weather data?	Y	N	N	N	N	Y	N	N	N	N	N
Method of data entry: H=Hand; F=data file memory; D=disk or tape	H,F,D	H,F,D	F,D	H,F	H	H,F,D	D	H,F	H,F,D	H	H,F
Number of chemicals hardwired (I=input by user)	—	960	10	32	16	5	—	1300	600	6	—
Source Emissions Models?											
Evaporation of Spilled Liquids?	N	Y	N	N	Y	N	N	Y	Y	Y	N
Flashing		Y			Y			Y	Y	Y	
Multicomponents		Y			Y			Y	Y	Y	
Entrainment as aerosols?		Y			Y			Y	Y	N	
Heat transfer, substrate to cloud?		Y			Y			Y	Y	Y	
Mass transfer in liquid phase?		Y			Y			Y	Y	Y	
Gas flux from container rupture?		Y			Y			Y	Y	Y	
Transport and Dispersion Model?											
Releases treated: I=Instantaneous; C=Continuous; V=Variable	I,C,V	I,C,V	C,V	I,C	I,C	I,C,V	I,C	I,C,V	I,C,V	I,C	I,C,V
Dense, Neutral, or Buoyant?	D,N,B	D,N,B	D,N,B	D,N,B	D,N	D,N,B	D,N,B	D,N	D,N,B	D,N,B	D
Jet	Y	N	Y	Y	Y	Y	Y	Y	Y	Y	Y
Evaporation of aerosols	Y	N	N	Y	Y	Y	Y	Y	Y	N	Y
Condens. of moisture in vapor cloud?	Y	N	N	N	Y	Y	Y	Y	Y	Y	N
Surface roughness?	Y	Y	N	N	Y	N	N	N	N	N	N
Complex terrain handled?	N	Y	N	Y	N	N	N	N	N	Y	N
U variations in time and space?	N	Y	N	Y	time	N	N	N	Y	Y	Y
Indoor concentrations?	N	Y	N	N	N	N	N	N	N	N	N
Building wake effects?	N	Y	N	Y	N	N	N	Y	N	N	N
Dispersion Model: B=Box or Slab; G=Gaussian; K=K/numerical	K	B,G	B,G	B	B,G	G	B	B,G,K	B,G	B,G	B
Along-wind dispersion?	Y	N	N	Y	N	Y	Y	Y	N	Y	Y
Lift-off?	N	Y	Y	Y	N	Y	Y	N	Y	Y	N
Chemical reactions in plume?	Y	Y	Y	Y	N	N	N	N	Y	N	N
Dry or wet deposition?	Y	Y	Y	Y	N	N	N	N	N	N	N
Concentration fluctuations?	Y	N	N	Y	Y	Y	N	Y	N	N	Y
Output											
Averaging time (minutes) (I=input)	—	—	10 min	—	—	—	—	—	—	—	10 min
Distance limits (km)	0–1000 km	—	0–10km	—	—	—	—	0–100 km	—	0–50 km	—

8

Evaluation of Models
with Field Data

All modelers evaluate their model with data to some extent, as can be seen in the model users guides or in their various reports and journal publications. These published evaluations range from simple scatter plots of observations versus predictions for one or two field tests to detailed comparisons involving statistical analysis of multiple sets of observations. However, very few model evaluation exercises involving several commonly used models have been carried out by independent groups (i.e., by persons not involved in the model development) using comprehensive sets of data. The results of independent model evaluations reported by Hanna et al. (1993) and Touma et al. (1995a) are described below, where the researchers evaluated whatever current version of the model was available. In most cases, the models have been updated in the intervening time period.

The evaluations reported in this section are all based on field experiments in the ambient (outside) environment. It should be pointed out that many laboratory data are also available for analysis, and several examples are discussed by Britter and McQuaid (1988), who make use of both field and laboratory data in deriving their correlations.

8.1. Description of Field Data Sets

The emphasis in this section is on field data sets rather than on laboratory data sets since the field data are more likely to reflect the physical processes occurring during actual accidental releases of hazardous chemicals. There are several independent field experiments in which useful data have been obtained over a variety of source scenarios and atmospheric conditions, and these experiments are considered below. Of course, many laboratory experiments have also been carried out in wind tunnels and water channels (see Britter and McQuaid (1988) for a summary of the laboratory experiments up to 1988), but they are not included in our analysis because some of these may be subject to scaling problems and other limitations and hence may not satisfactorily represent a full-scale atmospheric process.

Emphasis is on field data involving dense gas releases. About 20 or 30 such field experiments, each consisting of several chemical release trials, have been carried out since about 1970. Six of these dense gas field data sets (referred to by the code names Burro, Coyote, Desert Tortoise, Goldfish, Maplin Sands and Thorney Island) were selected by Hanna et al. (1993) for detailed evaluations. A subset of these data sets (Burro, Desert Tortoise,

TABLE 8-1
Summary of Characteristics of the Data Sets Used in Model Evaluations

	Burro	Coyote	Desert Tortoise	Goldfish	Maplin Sands	Thorney Island (Instantaneous)	Thorney Island (Continuous)
Number of Experiments	8	3	4	3	4, 8	9	2
Material	LNG	LNG	NH_3	HF	LNG,LPG	Freon & N_2	Freon & N_2
Type of Release	Boiling Liquid (dense gas)	Boiling Liquid (dense gas)	2-Phase Jet	2-Phase Jet	Boiling Liquid (dense gas)	Dense Gas	Dense Gas
Total Mass (kg)	10700–17300	6500–12700	10000–36800	3500–3800	LNG: 2000–6600 LPG: 1000–3800	4800	3150–8700
Duration (s)	79–190	65–98	126–381	125–360	60–360	Instant.	460
Surface	Water	Water	Soil	Soil	Water	Soil	Soil
Roughness (m)	.0002	.0002	.003	.003	.0003	.005–.018	.01
Stability Class	C–E	C–D	D–E	D	D	D–F	E–F
Max. Distance (m)	140–800	300–400	800*	3000	400–650	500–580	472
Min. Averaging Time (s)	1	1	1	66.6–88.3	3	0.06	30
Max. Averaging Time (s)	40–140	50–90	80–300	66.6–88.3	3	0.06	30
Reference	Koopman et al., 1982	Goldwire et al., 1983	Goldwire et al., 1985	Blewitt et al., 1987	Puttock et al., 1984	McQuaid & Roebuck, 1985	McQuaid & Roebuck, 1985

*Concentrations are measured beyond 800 m, but there are not well-instrumented measurement arcs at those distances.

and Goldfish) were used by Touma et al. (1995a) in their comparisons. An overview of the characteristics of the data sets is given in Table 8-1, which contains references for those readers who would like comprehensive details of the experiments. The data sets are described in more detail below.

Burro and Coyote. Both the Burro (Koopman et al., 1982) and Coyote (Goldwire et al., 1983) series of field trials were conducted in the desert over a broad dry lake bed at the Naval Weapons Center at China Lake, California. Sponsored by the U.S. Department of Energy and the Gas Research Institute, the trials consisted of releases of about 10,000 kg of LNG over a period of one to three minutes onto the surface of a 1 m deep pool of water, 58 m in diameter. The Burro series (8 trials) focused on the transport and diffusion of vapor from spills of LNG on water. In addition, the Coyote series (4 trials) expanded on the earlier Burro trials by including releases of liquefied methane and liquid nitrogen, and by studying the occurrence of rapid-phase-transitions (RPT) and of fires resulting from ignition of clouds from LNG spills. Note that "field trial" refers to a single experiment in which the dense gas was released over a period of a few minutes. Observations were made on monitoring arcs at distances ranging from 140 m to 800 m.

Desert Tortoise and Goldfish. These two sets of field experiments involved two-phase jets and were conducted over a desert surface at the Frenchman Flat area of the Nevada Test Site. However, because of an unusual rainstorm prior to the start of the Desert Tortoise experiments, the "dry" desert lake bed was covered by a shallow layer of water during most of the experiments. The first in the series, Desert Tortoise (Goldwire et al., 1985) was designed to document the transport and diffusion of ammonia vapor resulting from a cryogenic release of pressurized liquid ammonia. About 10,000 to 30,000 kg of pressurized liquid NH_3 was released over a time period of two to six minutes during four trials from a spill pipe pointing horizontally downwind at a height of about 1 m above the ground. The liquid jet flashed as it exited the pipe and its pressure decreased, resulting in about 18 percent of the liquid changing phase to become a gas. The remaining 82 percent of the NH_3-jet remained as a liquid, which was broken up into an aerosol by the turbulence inside the jet. Very little, if any, of the unflashed liquid was observed to form a pool on the ground. Dispersion of the vapor–aerosol cloud was dominated by the dynamics of the turbulent jet near the point of release, but the slumping and horizontal spreading of the cloud downwind of the jet zone indicated the influence of dense-gas dynamics at later stages. NH_3 concentrations and temperatures were obtained at elevations ranging from 1 to 8.5 m on towers located along arcs at distances of 100 m and 800 m downwind of the source. In most cases, nearly all of the plume was found to be located below the 6 m level of the towers and within the lateral domain of the towers. In addition, there were two arcs with up to eight portable ground-level stations at distances of 1400 m or 2800 m, and on occasion at 5500 m downwind. A diagram of the Frenchman Flat site is given in Figure 8-1, showing the locations of the meteorological tower and the concentration monitors relative to the source position (spill point).

The Goldfish trials (Blewitt et al., 1987) were carried out at the same field site and were very similar to the Desert Tortoise NH_3 trials described above. About 3500 kg of hydrogen fluoride (HF) was released over a time period of two to six minutes using a similar release mechanism and some of the same sets of instruments. A portion of the liquid HF flashed upon release during each of the three trials, creating a turbulent jet in which the unflashed liquid was broken up into an aerosol that remained in the jet cloud. No pooling of the liquid was observed on the ground. HF samplers were located at elevations of 1 to 8 m on cross-wind lines at distances of 300, 1000, and 3000 m from the source. In general the observed height of the HF cloud was less than the highest sampler level at the 300 m sampling line, but appeared to extend above the highest sample levels at the larger distances.

Maplin Sands. The 12 individual dispersion and combustion experiments conducted at Maplin Sands in 1980 (Puttock et al., 1984) involved the release of about 2000 to 6000 kg of liquified natural gas (LNG) or about 1000 to 3800 kg of refrigerated liquid propane (LPG) onto the surface of the sea over a time period of a few minutes. The size of a spill during each experimental trial was approximately 20 m^3. Because the objective of the trials was to study the behavior of LNG and LPG vapor clouds over the sea, the site was located on the tidal flats of the Thames estuary. Pontoons with either 4 m masts or 10 m masts were used to position meteorological instruments and 200 sampling instruments along arcs at distances from about 60 m to 600 m downwind of the spill area. A site diagram is given in Figure 8-2, showing the large numbers of monitors on broad concentric arcs.

Thorney Island. The Heavy Gas Dispersion Trials at Thorney Island (McQuaid and Roebuck, 1985) consisted of two types of source scenarios:

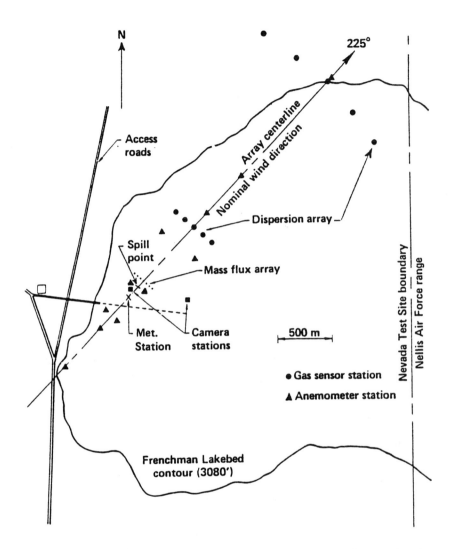

Figure 8-1. Site diagram for Desert Tortoise Series of dense gas dispersion experiments. (Goldwire et al., 1985). Concentration monitors are located on the three arcs labeled as "mass flux array" and "dispersion array."

(1) the instantaneous release of a preformed cloud of approximated 2000 m³ of dense gas over flat terrain (9 trials are included in the data set); and (2) the continuous release of 2000 m³ of heavy gas at a rate of about 5 m³/sec over about 6 or 7 minutes (2 trials are included in our data set). A cylindrical gas container with a volume of 2000 m³ was filled with a mixture of freon and nitrogen to simulate gases over a range of densities. For

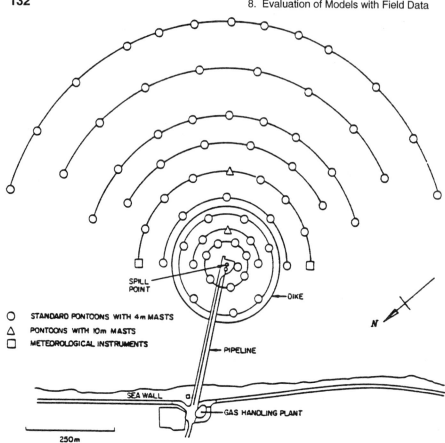

Figure 8-2. Site diagram for most of continuous LNG spills at Maplin Sands (Puttock et al., 1984). Concentration observations are made at the points marked by open circles or triangles.

instantaneous release trials, the sides of the gas container collapsed to the ground upon release. For continuous release trials, the gas container simply served as a storage tank. A 30 m tall meteorological tower with five levels of wind and temperature observations was located 150 m upwind from the release point. Thirty-eight towers were used to measure gas concentrations, with measurements taken at four levels at elevations from 0.4 m up to 14.5 m. The towers were placed on a rectangular grid with distances up to about 800 m from the release point.

Modelers' Data Archive (MDA). In order to carry out multiple runs of several dense gas models with the six databases in Table 8-1, which contain a total of 41 individual trials, the data were placed in computer files in a common format. A Modelers' Data Archive (MDA) was developed by

Hanna et al. (1993), using the criterion that it should contain only that information necessary for running and evaluating the models. Information is included that defines the experiment and trial, followed by a list of several chemical properties of the substance released. Chemical properties include the molecular weight, normal boiling point, latent heat of evaporation, heat capacity of the vapor phase, heat capacity of the liquid phase, and the density of the liquid. Physical properties of the release are then given, which not only provide specific dimensions, but also information on the general type of the release (source type and source phase) so that appropriate information can be passed to each of the models. Meteorological data appear next, such as wind speed and the Monin-Obukhov length scale, L. Site information in the MDA includes the surface roughness length, soil (or water) temperature, and a soil moisture indicator. The MDA file includes specific information related to the model application to a particular experiment and trial, such as concentration averaging time, concentration of interest (for specifying the lateral extent of the cloud or signaling how far downwind the model calculations should extend), receptor heights and distances, and observed maximum concentrations and lateral dispersion coefficients (σ_y) at each monitoring arc. The observed concentrations reported in these files represent the largest values measured along each monitoring arc at each distance, for each averaging time. Because the monitors on the arcs at each downwind distance could not cover every part of the plume, it is likely that the actual maximum concentration occurred at a location between the monitors, and therefore that the so-called observed maximum concentration is somewhat less than the true value. An intensive analysis of this problem was carried out, leading to the conclusion that, in most cases, there are errors of less than 10 to 20% introduced by this issue of incomplete coverage.

The Modelers' Data Archive and an accompanying descriptive report are available on diskette from Hanna et al. (1993).

8.2. Model Evaluation Procedures

The statistical evaluation procedures used in the Hanna et al. (1993) and Touma et al. (1995a) studies are similar and are described in articles by Hanna (1989) and Cox and Tikvart (1990). The software package used by Hanna et al. (1993) calculates the following model performance measures: the geometric mean bias (MG), geometric variance (VG), correlation coefficient (R), and fraction within a factor of two $(FAC2)$. Mathematical expressions for these performance measures are given below.

$$\text{Geometric Mean:} \quad MG = \exp(\overline{\ln C_O} - \overline{\ln C_p}) = \exp\left[\overline{\ln\left(\frac{C_O}{C_p}\right)}\right] \quad (8\text{-}1)$$

$$\text{Geometric Variance:} \quad VG = \exp[\overline{(\ln C_O - \ln C_p)^2}] = \exp\left[\overline{\left(\ln\left(\frac{C_O}{C_p}\right)\right)^2}\right]$$

$$(8\text{-}2)$$

$$\text{Correlation:} \quad R = \frac{\overline{(\ln C_O - \overline{\ln C_O})(\ln C_p - \overline{\ln C_p})}}{\sigma_{\ln C_p}\ \sigma_{\ln C_O}} \quad (8\text{-}3)$$

$$\text{Factor of Two:} \quad FAC2 = \text{Fraction of data for which } 0.5 \leq C_p/C_O \leq 2$$

$$(8\text{-}4)$$

where C_O is an observed concentration and C_p is the corresponding predicted concentration. An overbar indicates an average over the data set. The logarithmic forms of the mean bias and the scatter or variance [equations (8-1) and (8-2)] can be justified when there is wide range of magnitudes in the observed and predicted concentrations. If a data set were to contain concentrations that vary by less than a factor of two or three, then $\ln C$ can be replaced by C in equations (8-1) through (8-3), as done by Touma et al. (1995a) in their evaluations:

$$\text{Fractional Bias:} \quad FB = \frac{\overline{C_O} - \overline{C_p}}{0.5\ (\overline{C_O} + \overline{C_p})} \quad (8\text{-}5)$$

A "perfect" model would have MG, VG, R, and $FAC2$ equal to 1.0 and FB equal to 0.0. Geometric mean bias (MG) values of 0.5 to 2.0 can be thought of as "factor of two" overpredictions and underpredictions of the mean, respectively. A geometric variance (VG) value of about 1.6 indicates a typical factor of two scatter between the individual pairs of observed and predicted values. If there is only a mean bias in the predictions and no random scatter is present (e.g., if C_p always equals $2C_O$), then the relation, $(\ln VG) = (\ln MG)^2$, is valid, defining the minimum possible value of VG for a given MG.

The model evaluation software also allows the determination of whether the mean bias for a particular model is *significantly* different from zero, for example, at the 95% confidence level. It can also be determined whether

MG (or any other measure) for model A is significantly different from that for model B. For example, if model A has a geometric mean bias, *MG* = 1.1, and model B has *MG* = 1.3, then model A may appear to have a "better" *MG*, but this difference may not be significant at the 95% confidence level. The software employs bootstrap resampling methods (Efron, 1987) to estimate the standard deviation, σ, of the statistical performance measure in question. Then the 95 percent confidence intervals on the performance measure are calculated using the student-*t* procedure:

$$95\% \text{ confidence limits} = \text{mean} \pm t_{95}\sigma\left(\frac{n}{n-1}\right)^{\frac{1}{2}} \qquad (8\text{-}6)$$

So-called residual plots are also presented for each of the models and data sets, in order to determine if there are any trends in model overpredictions or underpredictions as a function of variables such as downwind distance, wind speed, stability, and observed concentrations. The so-called model residuals, or (ln C_p – ln C_O) = ln(C_p/C_O), are plotted on the *y*-axis in these graphs, and the independent variable is plotted on the *x*-axis. In order to avoid congestion of points in these graphs, the data within ranges of the independent variables are represented as "box plots," where the 2nd, 16th, 50th, 84th, and 98th percentiles (the mean ± one and two standard deviations for a Gaussian distribution) are shown.

8.3. Models Evaluated

There are dozens of models being marketed for application to calculating the transport and dispersion of hazardous chemical releases to the atmosphere. The characteristics of 22 of these were summarized in Chapter 7, using information as of April 1995. In 1991, Hanna et al. (1993) obtained several of the more commonly used models for testing, using the following criteria: (1) the model must run on a personal computer, (2) the code must be available to run independently (i.e., it was not desirable for the developers to make the runs), (3) the model should be applicable to all of the six field studies, (4) the model should be in use by several agencies and industries, and (5) the model should be different from the other models.

Hanna et al. (1993) selected three publicly available dense gas models (DEGADIS, HGSYSTEM, and SLAB), and six proprietary dense gas models (AIRTOX, CHARM, FOCUS, GASTAR, PHAST, and TRACE) for evaluations. The versions of these models were representative of April 1991, and are, in most cases, different from the April 1995, versions listed in Tables 7-1 and 7-2. For comparison purposes, the simple BM or Britter and McQuaid (1988) Workbook Model was also included. Touma et al. (1995a) selected a subset of this list of models (AIRTOX, DEGADIS, FOCUS, SLAB, TRACE) for their evaluation exercise, which considered model versions available in 1990. Brief overviews of these models and their reference documents are given below.

AIRTOX (Oct. 1990 Version) (Heinold et al. 1986 and Mills, 1988) is a proprietary model derived from established formulas in the literature. It calculates source emissions as well as dispersion.

BM (Britter and McQuaid, 1988). This "model" was originally a set of nomograms prepared using field and laboratory observations of dispersion of instantaneous and continuous releases of dense gases (see Section 5.6). Dimensionless variables were plotted on the nomograms, which were used by Hanna et al. (1993) to develop analytical formulas that best fit the curves drawn by hand by Britter and McQuaid (1988). This model does not calculate source emissions.

CHARM 6.1 (Radian, 1991) is unique among the proprietary models because it employs the puff superposition method for all types of releases. It calculates both source emissions and dispersion.

DEGADIS 2.1 (Havens and Spicer, 1985; Havens, 1988; Spicer and Havens, 1989) is a publicly available model in use by government agencies. It can handle dense or neutrally buoyant releases from point or area sources, and can simulate aerosol jets by input of a pseudo-plume density. Source emission rates must be calculated independently.

FOCUS 2.1 (Quest, 1990) is a part of a proprietary model that is a larger risk analysis package. It treats both source emissions and dispersion.

GASTAR 2.22 (Cambridge Environmental Research Consultants, 1990) is a multipurpose proprietary model, incorporating many effects, such as building downwash, not found in other models. It can calculate source emissions as well as dispersion.

HGSYSTEM (Nov. 1990 version) (Witlox et al. 1990, McFarlane et al. 1990, Witlox and McFarlane, 1994) can account for both aerosol jet sources and area sources and simulates the chemical reactions and thermodynamics of HF plumes. It calculates source emission rates for liquid spills and for some types of jet releases.

PHAST 2.01 (Technica, 1989) is a proprietary model that is part of a larger risk analysis package, and can calculate both source emissions and dispersion.

SLAB (Ermak, 1990) is a general-purpose publicly available model that applies to all types of accidental releases (point or area sources, dense or neutrally buoyant gases, continuous or instantaneous releases). Its development was sponsored by the Department of Energy (DOE). It can simulate aerosols by means of empirical approximations. Source emission rates must be calculated independently.

TRACE 2.54 (DuPont, 1989) was originally part of a real-time emergency response system. Like the other proprietary models, it handles both source emissions and dispersion.

Although the six proprietary models can all simulate the source emission rate, this asset was sometimes found to be a liability in the model evaluation exercise. The problem is that the models may sometimes calculate a mass emission rate that does not agree with the observed value, which may naturally cause the predicted concentrations to be biased by at least the same amount. In contrast, the other (publicly available) models were provided with source emission rates that agree closely with the true value observed during the field experiments. It was found that, for example, the AIRTOX model's underpredictions of concentrations for the Maplin Sands field data are due primarily to underpredictions of the source emission rate for LNG and LPG spills over water. However, it had been decided that the models should all be run in the mode suggested in their user's guide, and therefore it was not appropriate to "adjust" their predicted source emission rates.

As mentioned earlier, most of the models evaluated here have been updated in the past few years. The relative performance of the updated model versions may differ from that reported below.

8.4. Results of Model Evaluations

The ability of the models to simulate maximum plume centerline concentrations for short-term averages was evaluated by Hanna et al. (1993) using the statistical tests described in Section 8.2. In all cases, the model predictions at ground elevation are used for comparisons, since only a few models allow the actual receptor height (usually about 1 m) to be input. Graphs of the geometric mean (i.e., the relative mean bias), MG, versus the geometric variance (i.e., the relative variance), VG, for concentration predictions for the two groups of field data (continuous dense gas releases and instantaneous dense gas releases) are given in Figures 8-3 and 8-4, respectively. A perfect model compared against perfect observations would be placed at the $MG = 1$ and $VG = 1$ point on this figure. Lines extending

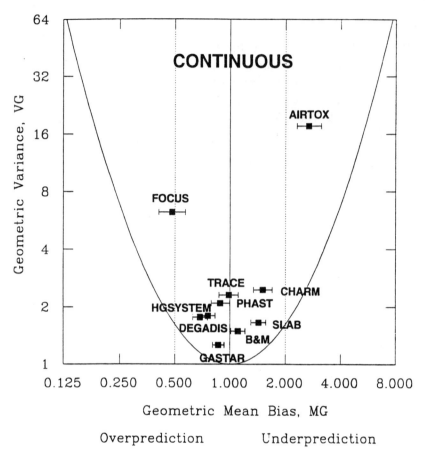

Figure 8-3. Model (1991 versions) performance measures, geometric mean bias MG – $\exp\overline{(\ln C_o - \ln C_p)}$ and geometric variance $VG = \exp\overline{[(\ln C_o - \ln C_p)^2]}$ for maximum plume centerline concentration predictions and observations. 95% confidence intervals on MG are indicated by the horizontal lines. The solid parabola is the "minimum VG" curve. The vertical dotted lines represent "factor of two" agreement between mean predictions and observations. For continuous dense gas data sets (Burro, Coyote, Desert Tortoise, Goldfish, Maplin Sands and Thorney Island), involving a total of 32 trials and 123 points for the shortest available instrument averaging times (from Hanna et al, 1993).

horizontally from each point indicate 95% confidence limits on the geometric mean, MG. A model that has no random scatter but suffers a mean bias would be placed somewhere along the parabolic curve, $(\ln VG)^{1/2} = \ln MG$, which represents the minimum possible value of VG that corresponds to a particular MG. Therefore, all of the points must lie within or above the parabola. Furthermore, the vertical dotted lines on the figures mark the

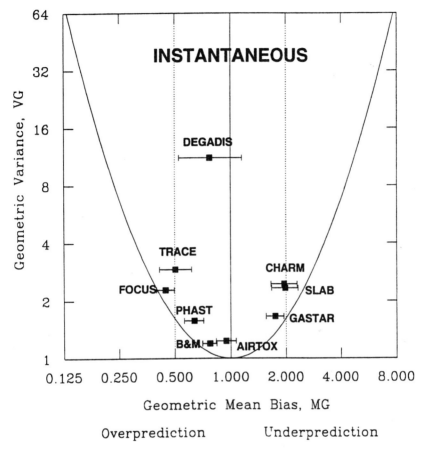

Figure 8-4. Model (1991 versions) performance measures, geometric mean bias $MG = \exp(\overline{\ln C_o - \ln C_p})$ and geometric variance $VG = \exp[\overline{(\ln C_o - \ln C_p)^2}]$, for maximum plume centerline concentration predictions and observations. 95% confidence intervals on MG are indicated by the horizontal lines. The solid parabola is the "minimum VG" curve. The vertical dotted lines represent "factor of two" agreement between mean predictions and observations. For the instantaneous dense gas data set (Thorney Island), involving a total of 9 trials and 61 points (from Hanna et al., 1993).

values of MG that correspond to "factor-of-two" differences in the means. Models that fall between the vertical dotted lines produce estimates that are within a factor of two of observed values, on average.

The two figures show that no one model produces consistently good performance. Nevertheless, there is a cluster of models that produce reasonable performance on each figure. Also, there are one or two models on each figure that produce results that are clearly poorer than the other models. Retrospective analyses of these results can uncover the reasons for poor performance in some cases. For example, the underpredictions by the AIRTOX model are found to be due to limitations in the source characterization algorithm, and not to problems in the transport and dispersion algorithms. The AIRTOX source algorithms do not satisfactorily handle the Maplin Sands release scenarios involving spills of LNG and LPG on water surfaces, since the maximum AIRTOX conductivity coefficient is for wet soils.

The results for continuous dense gas data sets illustrated in Figure 8-3 include all trials and all monitoring arcs for short-term averages for the data sets from quasi-continuous releases of dense-gas clouds (Burro, Coyote, Desert Tortoise, Goldfish, Maplin Sands, and Thorney Island [continuous]). The geometric mean bias (MG) values for all of the models except FOCUS and AIRTOX are within the dotted vertical lines, indicating that, on average, peak modeled concentrations are within a factor of two of peak observed concentrations.

The FOCUS and AIRTOX models have a relatively large geometric variance, VG. As mentioned earlier, the AIRTOX model underpredictions are due primarily to its inability to simulate heat transfer from underlying water surfaces. The other nine models are "bunched" within a VG range of about 1.4 to 2.6. GASTAR has the smallest VG, indicating a typical scatter of slightly less than a factor of two. Note that the model (TRACE) with the best geometric mean does not have the smallest variance, and the model (GASTAR) with the smallest variance does not have the best geometric mean.

It can be concluded from Figure 8-3 that eight models (B&M, CHARM, DEGADIS, GASTAR, HGSYSTEM, PHAST, SLAB, and TRACE) have fairly good performance, with mean biases within a factor of two and a scatter of about a factor of two.

Figure 8-4 shows the results for the instantaneous dense gas releases at Thorney Island. It is seen that the scatter or variance for all models (except for DEGADIS) tends to be dominated by the mean bias. For this data set, the AIRTOX, B&M, and PHAST models show fairly good performances, with mean bias and scatter much less than a factor of two. The other models tend to have mean biases and scatter equal to a factor of two or greater.

An interesting result on Figures 8-3 and 8-4 is the relatively good performance of the Britter and McQuaid (BM) benchmark model. The simple Britter and McQuaid (1988) model is able to match the observations because it is an empirical relation derived from laboratory data and from some of the same field data that it is being tested against.

The evaluations reported by Touma et al. (1995a) of AIRTOX, DE-GADIS, FOCUS, SLAB, and TRACE with the Burro, Desert Tortoise, and Goldfish data sets emphasized the fractional bias (*FB*) performance measure. Table 8-2 contains values of *FB* for the various models and data sets, where "far" and "near" refer to the relative locations of the monitoring arcs. It is seen that the "far" and "near" statistics show no major trends and that all models perform within a factor of two for the combined data sets. The sign of *FB* in Table 8-2 is consistent with the findings by Hanna et al. (1993) in Figure 8-3 (e.g., SLAB tends to underpredict while DEGADIS tends to overpredict). Touma et al. (1995a) conclude on the basis of their analysis of several performance measures for concentrations and plume widths that "model performance varied and no model exhibited consistently good performance across all three data bases."

Residual plots of $\ln(C_p/C_O)$ versus variables such as wind speed were made for all models by Hanna et al. (1993). An example of a set of residual plots for one model (HGSYSTEM) is given in Figure 8-5 for the continuous dense gas data set. The downwind distance (x), wind speed (u), and Pasquill-Gifford stability class (*PG*) are used as independent variables. There is seen to be a slight trend from overpredictions of 30 to 40% at the shortest downwind distances (100 m) to underpredictions of about the same amount at the longest downwind distances (800 m). There are no clear trends with wind speed or stability class.

Predictions of plume widths are also important for estimating the size of zones of hazardous gas impacts. Observations and predictions of plume widths are compared in Figure 8-6 for the continuous dense gas data. It should be noted that not all models produce outputs of plume width and not all field data sets allowed plume widths to be estimated from the monitoring data. Because the predicted lateral distributions were often non-Gaussian, the predicted plume half-width, $W/2$, was defined as the lateral distance where the concentration dropped to $e^{-1/2}$ of the centerline value. For a Gaussian distribution, $W/2$ is exactly equal to the standard deviation, σ_y, of the lateral concentration distribution.

The vertical geometric variance (*VG*) scales for widths on Figure 8-6 range from 1.0 to 4.0, whereas the scales for concentrations on Figure 8-3 range from 1.0 to 64.0. The scales have been adjusted so the points fit comfortably on the figure. Comparing Figure 8-3 with Figure 8-6, it is immediately evident that predictions of half-widths, $W/2$, are generally more successful, overall, than are the predictions of centerline concentration. The largest values of variance for the cloud widths are much smaller than the variances for the centerline concentrations, probably due to the smaller range

TABLE 8-2
Average Fractional Bias, $FB = (\overline{C_o} - \overline{C_p})/[0.5(\overline{C_o} + \overline{C_p})]$, Using Maximum
Observed and Predicted Concentration Values*

Distance	Model	Desert Tortoise	Goldfish	Burro	Combined
			Experiment		
Far	AIRTOX	0.12	1.21	0.52	0.62
	DEGADIS	−0.25	0.37	−1.15	−0.34
	FOCUS	−0.14	1.07	−1.88	−0.32
	SLAB	0.89	0.93	−0.01	0.61
	TRACE	−0.01	0.51	−1.60	−0.37
Near	AIRTOX	−1.39	1.16	0.76	0.18
	DEGADIS	−1.55	0.47	−1.24	−0.77
	FOCUS	−0.90	0.19	−1.78	−0.83
	SLAB	0.87	1.10	0.31	0.76
	TRACE	−0.30	0.71	−0.81	−0.13
Combined	AIRTOX	−0.63	1.19	0.64	0.40
	DEGADIS	−0.90	0.42	−1.20	−0.56
	FOCUS	−0.52	0.63	−1.83	−0.57
	SLAB	0.88	1.01	0.15	0.68
	TRACE	−0.16	0.61	−1.21	−0.25

*From Touma et al. (1995a). A negative value of FB indicates an **over**prediction. These results are for
the 1990 versions of the models.

of observed values of cloud widths. Furthermore, the variations in variance
are largely due to variations in mean bias, as expressed by the curve ($\ln VG)^{1/2} = \ln MG$.

The 1991 versions of the models tend to overpredict the plume widths
for the dense-gas releases in Figure 8-6. AIRTOX, HGSYSTEM, PHAST,
and SLAB overpredict the widths by less than about 50 percent, on the
average. DEGADIS and GASTAR, overpredict the width by a factor of two
or more. The tendency of the models to overpredict dense gas plume widths
is found to be connected with a tendency to underpredict dense gas plume
depths by about the same amount. Consequently, the models tend to predict
dense gas clouds that are broader and shallower than observed.

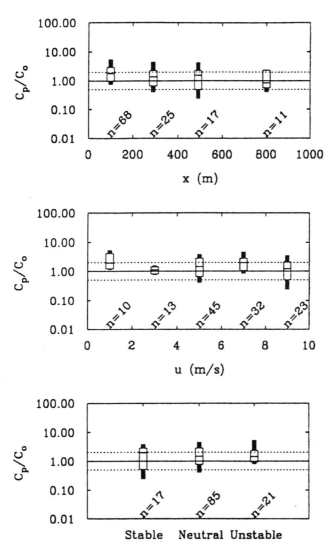

Figure 8-5. Distributions of model residuals, C_p/C_o, for HGSYSTEM (1991 version) for maximum concentrations on the plume centerline. Independent variables are downwind distance x (top figure), wind speed u (middle figure), and Pasquill-Gifford stability class PG (bottom figure). The "box plot" format indicates the 2nd, 16th, 50th, 84th and 98th percentiles of the cumulative distribution function of the n points in the box. For continuous dense gas data set (Burro, Coyote, Desert Tortoise, Goldfish, Maplin Sands and Thorney Island). From Hanna et al. (1993).

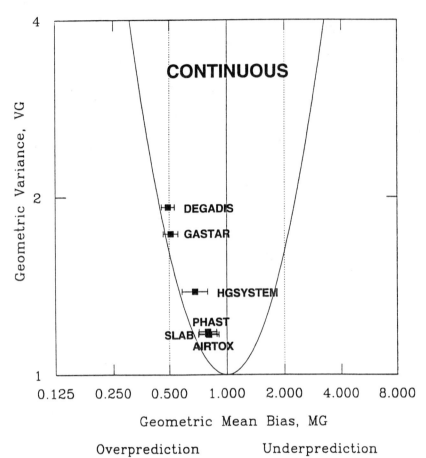

Geometric Mean Bias, MG

Overprediction   Underprediction

Figure 8-6. Model performance measures, geometric mean bias $MG = \exp(\overline{\ln W_o - \ln W_p})$ and geometric variance $VG = \exp[\overline{(\ln W_o - \ln W_p)^2}]$, for plume width predictions and observations. 95% confidence intervals on MG are indicated by the horizontal lines. The solid parabola is the "minimum VG" curve. The vertical dotted lines represent "factor of two" agreement between mean predictions and observations. For continuous dense gas data sets (Burro, Coyote, Desert Tortoise and Goldfish), involving 18 trials and 30 points. From Hanna et al. (1993).

It was concluded by Hanna et al. (1993) that the dense gas models (1991 versions) that produced the most consistent predictions of plume centerline concentrations across the dense gas data sets are the Britter and McQuaid, CHARM, GASTAR, HGSYSTEM, PHAST, SLAB, and TRACE models, with relative mean biases of about ± 30% or less and magnitudes of relative scatter that are about equal to the mean. The dense gas models (1991 versions) tended to overpredict the plume widths and underpredict the plume depths by about a factor of two.

Although it is not shown in the preceding figures, the simple Gaussian plume model (see Section 5.4) for nondense gases was also run for these data sets. It was found that the centerline concentrations predictions of the Gaussian plume model matched the observations as well as any dense gas model, although the Gaussian plume model underpredicted the dense cloud widths and overpredicted the dense cloud heights by factors of three or four or more.

Evidently, even though the geometry of a dense gas plume is quite different than the geometry of a passive gas plume, there is a compensating effect that results in the plume cross-sectional area and hence the centerline concentration remaining nearly the same (assuming other conditions such as mass emission rate and meteorology are the same).

These model evaluation results are based on 1990 or 1991 versions of the models, and are included here in order to illustrate the types of results that can be generated. Since most models have been updated, the evaluation results would be likely to change of these performance measures were recalculated using the latest model versions.

9

Summary of
Seven Worked Examples

Appendices A through G contain complete descriptions of the applications
of specific models to seven worked examples, including discussions of how
to set up input files, examples of tabulated and plotted model predications
of cloud centerline concentrations and cloud widths, and interpretation of
the results. All input and output files for the various models that are applied
to the scenarios are contained in a diskette inserted in a pocket in the inside
back cover. This section provides a summary description of the seven
scenarios and a tabulation of the results.

9.1. Description of Seven Scenarios Used for Worked Examples

In order to demonstrate the use of vapor cloud dispersion models for specific
applications, seven typical scenarios have been proposed, covering a range
of chemicals, storage conditions, and accidental release conditions. These
seven scenarios should cover most of the types of accidental releases that
would be encountered by the modeler. The scenarios have been defined so
that some of the source emissions formulas presented in Section 4 could be
used, and so that several of the publicly available transport and dispersion
models available in April, 1995, and discussed in Sections 7 and 8 could
be tested (ALOHA, DEGADIS, HGSYSTEM, INPUFF, ISC, and SLAB).
Schematic diagrams of the seven release scenarios are given in Figure 9-1
and brief descriptions of the scenarios and the models that were applied are
listed in Table 9-1. Note that one base run and three sensitivity runs were
made with each model in each scenario, in order to compare the sensitivities

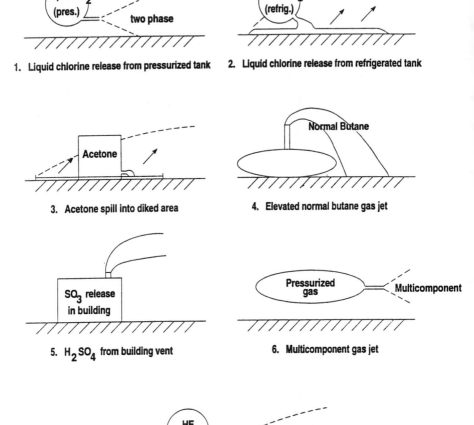

Figure 9-1. Schematic diagram of release scenarios for seven worked examples.

of predicted concentrations to variations in some of the input parameters. In addition, a Scenarios Data Archive (SDA) has been set up, containing all input data sufficient to run the listed models (see Table 9-2). The SDA is included in the diskette in the inside back cover. With this SDA and additional interface software for various vapor cloud dispersion models, it was possible to change one or two input parameters and relatively quickly rerun all models for all scenarios by striking only a few keys on a computer keyboard.

TABLE 9-1
Overview of Seven Scenarios Used for Worked Examples,
Including Models to Be Tested and Sensitivity Runs

Scenario	Brief Descriptions	Source Emissions Models	Dispersion Models	Input Varied in Sensitivity Runs
1	Continuous pressurized horizontal release of liquid chlorine	Two-phase pressurized jet formula (Fauske and Epstein)	DEGADIS HGSYSTEM SLAB	Stability/wind speed Surface roughness Source diameter
2	Evaporation from refrigerated liquid spill of chlorine	Evaporation models (ALOHA, LPOOL)	DEGADIS HGSYSTEM SLAB	Stability/wind speed Surface roughness Dike diameter
3	Liquid spill of large tank of acetone	Slowly evaporating pool formula (Fleischer) and ALOHA	ALOHA DEGADIS HGSYSTEM	Stability/wind speed Averaging time Dike diameter
4	Continuous vertical dense normal- butane gas jet	Prescribed source term	DEGADIS SLAB	Stack height Source diameter Averaging time
5	Slowly time-varying venting of warm sulfuric acid plume from building	Formulas for chemical reactions and thermodynamics of SO_3 reaction with water vapor	INPUFF ISC	Stability/wind speed Building height Stack height
6	Pressurized gaseous release of multicomponent mixture	Prescribed source term	ALOHA HGSYSTEM SLAB	Stability/wind speed Averaging time Increase source term
7	Transient (mitigated) area-source release of HF	Prescribed source term	DEGADIS HGSYSTEM	Relative humidity Mitigation start time Surface roughness

The Scenarios Data Archive (SDA) in Table 9-2 contains several lines of "header" information, including the scenario number and description, the models to be applied, and the name of the chemical. The following specific information is then given.

Properties of Released Material

The chemical and physical properties of the released material are included in the SDA in the first block of data, containing values for nine variables.

TABLE 9-2 Scenarios Data Archive (SDA) for Worked Examples, Containing All Input Data Sufficient to Run Models Listed in Table 9-1

	Scenario 1	Scenario 2	Scenario 3	Scenario 4	Scenario 5	Scenario 6	Scenario 7
SCENARIO DESCRIPTION	Continuous pressurized horizontal release of liquid chlorine	Continuous liquid (cryogenic spill of large tank of chlorine)	Continuous liquid (noncryogenic spill of large tank of acetone)	Continuous vertical dense gas jet of normal butane	Slow time-variable venting of sulfuric acid from building (passive dispersion)	Pressurized gaseous release of multi-component mixture of hydrocarbons	Transient (mitigated) area-source release of HF
MODELS TO BE APPLIED	DEGADIS, HGSYSTEM, SLAB	DEGADIS, HGSYSTEM, SLAB	ALOHA, DEGADIS, HGSYSTEM	DEGADIS, SLAB	INPUFF, ISC	ALOHA, HGSYSTEM, SLAB	DEGADIS, HGSYSTEM
PROPERTIES OF RELEASED MATERIAL							
Name	Chlorine (Cl$_2$)	Chlorine (Cl$_2$)	Acetone (C$_3$H$_6$O)	Normal Butane (nC$_4$H$_{10}$)	Sulfuric Acid (H$_2$SO$_4$)	Major component: Ethane (C$_2$H$_6$)[a]	Hydrogen Fluoride (HF)
Molecular weight (kg/kmole)	70.91	70.91	58.1	58.12	98.08	30.07	20.01
Normal boiling point (K)	239.1	239.1	329.4	272.7	n/a	184.52	292.7
Latent heat of evaporation (J/kg)	287,840	287,840	545,000	363,000	n/a	489,360	373,000
Specific heat at constant pressure for vapor (J/kg/K)	498	498	1467	1715	n/a	1774	1450
Specific heat of constant pressure for liquid (J/kg/K)	926	926	2070	2054	n/a	2346	2528
Density of liquid (kg/m^3)	1574	1574	788	623	n/a	547	987
Molecular diffusivity (m^2/s)	1.22 x 10^{-5}	1.22 x 10^{-5}	1.23 x 10^{-5}	n/a	n/a	n/a	n/a
Kinematic viscosity (m^2/s)	4.48 x 10^{-6}	4.48 x 10^{-6}	1.10 x 10^{-5}	n/a	n/a	n/a	n/a
Mole fraction (%) in pollutant mix	100	100	100	100	100	60	100
SOURCE CONFIGURATION							
Storage temperature (K)	298	239.1	303	303	293	293	308
Release temperature (K)	239.1	239.1	303	303	293	293	292.7
Storage pressure (atm)	8	1	1	1	1	4	10
Storage phase	liquid	liquid	liquid	gaseous	gaseous	gaseous	gaseous
Release phase	2-phase	liquid	liquid	gaseous	gaseous	gaseous	2-phase
Rupture/stack/source diameter (m)	0.0381	0.0381	0.0508	0.75	1	0.1	24
Rupture/stack/source height (m)	1.5	1.5	1	5	20	1	0
Mass in tank (kg)	80,000	80,000	143,743	n/a	n/a	n/a	n/a
Tank diamter (m)	4	4	7.62	n/a	n/a	n/a	n/a
Tank height (m)	5	5	7.62	n/a	n/a	n/a	n/a
Fluid level (m)	4.04	4.04	4	n/a	n/a	n/a	n/a

Pipe length (m) for pipe rupture	3.81	n/a	n/a	n/a	n/a	n/a	n/a
Dike, centered at source, diam. (m)	n/a	n/a	21.9	n/a	n/a	n/a	n/a
Dike height (m)	n/a	n/a	0.914	n/a	n/a	n/a	n/a
Ground surface composition	n/a	concrete	concrete	n/a	concrete	n/a	n/a
Building height (m)	n/a	n/a	n/a	n/a	18	n/a	n/a
Building width (m)	n/a	n/a	n/a	n/a	22.36	n/a	n/a
Mass fraction that flashes to vapor	0.189	n/a	n/a	n/a	n/a	n/a	0.104
Vapor/aerosol mix density (kg/m³)	18.9	n/a	n/a	n/a	n/a	n/a	7.98
Source Richardson number	1.36×10^6	782	3.61	28100	n/a	106	957
Richardson number for down-wind effects for point source	3.45×10^4	n/a	n/a	4220	n/a	10.6	n/a
SOURCE STRENGTH	11.04	9.3 (pool radius = 20 m)	2.34 (effective radius = 10.27 m)	15	0.272 exponentially decaying to 0.044 over 1 hour	5	10 for first 60 sec. 1 for next 540 sec.
Type	steady-state	steady-state	ALOHA: transient HGSYSTEM and DEGADIS: steady-state	steady-state	INPUFF: transient ISC: pseudo-steady-state	steady-state	transient
Method of deactivation	calc. from above source config, (Fauske & Epstein)	calc. by LPOOL module of HGSYSTEM (ALOHA yields similar evap. rate but locks pool radius)	calc. from above source config, for DEGADIS (Fleischer)	direct user-input	calc. from an assumed building exchange rate	direct user-input	direct user-input
METEOROLOGICAL CONDITIONS							
Temperature (k)	298	298	303	303	300	303	293
Measuring height of temp. (m)	10	10	10	10	10	10	10
Wind speed (m/s)	2	2	5	2	2	5	2
Measuring height of wind speed (m)	10	10	10	10	10	10	10
Stability class	F	F	D	F	C	D	F
Pressure (atm)	1	1	1	1	1	1	1
Relative humidity (%)	80	80	50	50	50	75	90
Monin–Obukhov length (m)	12.2	12.2	9999	14.3	−74.2	9999	14.3
Friction velocity (m/s) profile method	0.0808	0.0808	0.344	0.0986	0.303	0.511	0.0986
Wind speed at 1 m (m/s) (profile method)	0.791	0.791	3.02	0.654	0.486	2.06	0.654

Multicomponent mixture: 60% (mole) ethane, 20% (mole) nitrogen, 10% (mole) propane, 10% (mole) oxygen. Weighted MW = 31.25 kg/mole. Weighted cp = 47.29 J/mole/K = 1513 J/kg/K.

TABLE 9-2 (Continued)

	Scenario 1	Scenario 2	Scenario 3	Scenario 4	Scenario 5	Scenario 6	Scenario 7
SCENARIO DESCRIPTION	Continuous pressurized horizontal release of liquid chlorine	Continuous liquid (cryogenic spill of large tank of chlorine	Continuous liquid (noncryogenic spill of large tank of acetone	Continuous vertical dense gas jet of normal butane	Slow time-variable venting of sulfuric acid from building (passive dispersion)	Pressurized gaseous release of multi-component mixture of hydrocarbons	Transient (mitigated) area-source release of HF
SITE CONDITIONS (for dispersion calculations)							
Surface roughness (m)	0.03	0.03	0.03	0.1	0.5	0.2	0.1
Surface type	uncut grass	uncut grass	uncut grass	industrial	town outskirts	industrial	farmland
CONCENTRATION INFORMATION							
Averaging time (s)	600	600	600	600	600	600	600
IDLH (ppm)	30	30	20,000	10,000	20	50,000	30
UEL (ppm)	n/a	n/a	128,000	84,100	n/a	125,000	n/a
LEL (ppm)	n/a	n/a	26,000	18,600	n/a	30,000	n/a
SENSITIVITY RUNS							
Run 1	Double source pipe diameter to 0.0762 m, thus increasing source strength by a factor of 4	Change stability to class D (neutral) and increase wind speed to 10 m/s	Change stability to class F (stable) and decrease wind speed to 2 m/s	Increase stack height from 5 m to 20 m	Change stability to class F (stable)	Change stability to class F (stable) and decrease wind speed to 2 m/s	Change relative humidity from 90% to 25%
Run 2	Change stability to class D (neutral) and increase wind speed to 5 m/s	Dike with radius of 5 m and and height of 0.914 m (effective pool radius = 21.57 m and rate = 0.8 kg/s)	Double dike diameter from 21.9 m to 43.8 m (effective pool radius = 21.57 m and rate = 8.65 kg/s)	Decrease stack diameter from 0.75 m to 0.30 m	Increase building height from 18 m to 36 m; increase stack vent height from 20 m to 38 m	Increase source strength by a factor of 4	Extend initial 10 kg/s source term from 60 sec to 300 sec
Run 3	Increase roughness length by one order of magnitude to $z_0 = 0.3$ m	Increase roughness length by one order of magnitude to $z_0 = 0.3$ m	Change averaging time from 600 sec to 0 sec	Change averaging time from 600 sec to 0 sec	Increase stack vent height from 20 m to 40 m	Change averaging time from 600 sec to 0 sec	Increase surface roughness z_0 from 0.1 to 0.5 m

It is assumed that properties such as the specific heat, liquid density, molecular diffusivity, and kinematic viscosity do not vary with temperature. Nearly all of these variables for various gases can be found in standard sources such as Perry et al. (1984). Although molecular diffusivities are usually not extensively tabulated, they can be readily calculated by the Lennard-Jones functions (e.g., Bird et al., 1960).

All scenarios involve releases of a single chemical, except for Scenario 6, where a multicomponent mixture of ethane, propane, nitrogen, and oxygen is assumed. The molar fraction for each component in Scenario 6 is included in the table. However, the listed properties for Scenario 6 are for ethane, the dominant component. Since that scenario involves a gaseous release with no phase changes, several mixture properties such as the boiling point, the specific heat of the liquid, the density of the liquid, the molecular diffusivity, and the kinematic viscosity are not relevant and therefore are not listed. The molecular weight and specific heat of the gas are important for the solution to Scenario 6, and are included in parentheses on a separate line of the table (calculated by a mass-weighted average over the components). Also it is noted that the properties of H_2SO_4 released in Scenario 5 are not required since it is assumed to be a passive "tracer" release at an elevation equal to the vent height plus the plume rise.

Source Configuration

This section of the SDA contains 21 entries concerning the characteristics of the source, including the storage temperature and pressure, the storage and release phase of the chemical, and the rupture diameter and elevation. Notice that the two releases of pressurized liquids (i.e., Cl_2 and HF in Scenarios 1 and 7) result in two-phase jets. Although the initial release phase for the two evaporating pools (i.e., Cl_2 and acetone in Scenarios 2 and 3) is liquid, the source term for the gas entering the atmosphere is determined by the evaporation rate from the liquid pool spilled on the ground. For these two scenarios, the rupture diameter and the characteristics of the liquid in the tank are used to calculate the liquid flow rate into a pool which then determines the pool diameter, which is in turn used to calculate the evaporation rate for input to the dispersion models.

In addition, this section of the SDA contains data describing the release mechanisms for Scenarios 1, 2, and 3. The source emission rates are functions of the tank geometry (an upright cylindrical shape is assumed), the amount of chemical stored in the tank, the size and elevation of the rupture, the length of the pipe if a pipe rupture is involved, the dike diameter, and the ground surface composition. The composition of ground surface is required to estimate the evaporation rate for the liquid pool cases (Scenarios 2 and 3).

Data regarding the release mechanisms for the remaining Scenarios (4 through 7) are not provided in the SDA, since the emission rate is directly specified by the user in Scenarios 4, 6, and 7, and is calculated based on chemical reactions, thermodynamics, and air exchange rates within the building in Scenario 5, as described in detail in Appendix E.

Building dimensions used as input to a building downwash algorithm in the ISC model are specified for Scenario 5.

This section of the SDA also lists some other parameters that must be derived as part of the source term calculation, such as the mass fraction that flashes to vapor, the density of the mixture of vapor and aerosols for two-phase releases (Scenarios 1 and 7, see Appendices A and F), and the source Richardson number as defined by equations (5-1) through (5-4).

Source Strength

The source strength listed in the top few lines of the second page of Table 9-2 is calculated using the approaches described in words in the table. The source term for Scenarios 1, 2 and 3 is estimated from the procedures described in Section 4, and is directly specified by the user for Scenarios 4, 6, and 7. The source term in Scenario 5 for the in-building release of H_2SO_4 is calculated using analytical approximations described in detail in Appendix E. For the boiling chlorine pool in Scenario 2, the methodology described in Section 4 to calculate the source term cannot be easily carried out by hand, since the solutions vary with time. Thus, for that scenario, the LPOOL module of the HGSYSTEM model was used to estimate the source term and the average value over the first hour after release was assumed to be valid. For Scenarios 2 and 3, the liquid pool radius was also specified in the SDA, since the evaporation occurs across that radius and is needed to initialize the dispersion model.

The release type (i.e., steady-state versus transient) for each scenario and model is also included under this section of the SDA. The Appendices provide the detailed source term calculations for each scenario, if applicable.

Meteorological Conditions

The meteorological conditions for each scenario are specified as ten data entries in this section of the SDA on the second page of Table 9.2. The first seven parameters (temperature and its measuring height, wind speed and its measuring height, stability class, pressure, and relative humidity) are directly input by the user. The Monin-Obukhov length, L, has been independently estimated for these scenarios using Golder's (1972) nomogram based on knowledge of stability class and surface roughness length. Most models allow the option of direct input of L or calculating L internally. The last two entries in this section of the SDA, the friction velocity, u_*, and the wind speed at 1 m, u_1, are calculated from the well-known profile method based on Monin-Obukhov similarity theory as outlined in Section 3.3 [see equation (3-4)]. The wind speed at 1 m, u_1, is listed because it is representative of the initial advection speed of the cloud for most of the scenarios. The values of L, u_*, and u_1 are listed in the SDA mainly for reference. The friction velocity, u_*, is used later for calculating the source Richardson number. These three parameters were usually not used to run the dispersion models, since most models were instructed to calculate L, u_*, and cloud advection speed internally.

Site Conditions

There are two entries in the SDA for site conditions—the surface roughness, z_0, and the surface type. Note that the surface type listed here is only a verbal description of z_0, whereas the surface type listed under the "Source Configuration" section is directly used to parametrize the effects of the ground heat flux on the evaporation rate from a liquid pool.

Concentrations of Interest

The next section of the SDA contains four entries related to concentrations of interest and their averaging time. It is assumed in the first set of runs (i.e., the base runs) that the concentration averaging time, T_a, for all scenarios is 600 seconds. Sensitivity runs are carried out for small averaging times which are more appropriate for determining the flammable effects of the cloud. All heavy gas dispersion models considered here, except for ALOHA, allow the user to specify the concentration averaging time. The ISC passive gas dispersion model assumes an averaging time of one hour. The INPUFF passive gas dispersion model allows the user to specify the concentration averaging time.

The IDLH (Immediately Dangerous to Life or Health) concentration is listed for all chemicals, and the UEL (Upper Explosive Limit) and the LEL (Lower Explosive Limit) concentrations are listed for some chemicals. The

UEL and LEL are not listed if flammability is not a major concern for the chemical. The values of IDLH, UEL, and LEL concentrations for chlorine, acetone, sulfuric acid, and hydrogen fluoride were obtained from the handbook published by NIOSH (1987). The values of UEL and LEL concentrations for normal butane and ethane were obtained from a document prepared by the EPA (1990). The IDLH concentrations for normal butane and ethane are taken as the "no effect level for man" concentration limits suggested by Verschueren (1983).

Sensitivity Runs

The last section of the SDA contains information concerning input conditions for the three sensitivity runs, including only those conditions changed from the base run. Three sensitivity runs were carried out for each model for each scenario. One or two input parameters were varied for each sensitivity run. Most of the time the changes included stability class/wind speed, averaging time T_a, or surface roughness z_0. Sometimes the stack height or the pipe or dike diameter were changed. In Scenario 6 the building size was changed, and in Scenario 7 the mitigation timing was changed.

9.2. Overview of Predicted Concentrations

Appendices A through G contain detailed discussions of the model predictions for each scenario, including several tables and graphs. Examples of the results are given in Table 9-3, where all predicted cloud centerline concentrations (at ground-level) are listed for downwind distances ranging from 100 m to 2000 m. Predictions for all models and all sensitivity runs are listed.

It is not practical here to try to fully analyze the 432 concentrations listed in Table 9-3. The appendices provide discussions of many aspects of these results. Nevertheless, it is important to point out that Table 9-3 shows that there are often differences among the model predictions for a given scenario at a given distance. Agreements as close as ±10% or 20% between models occasionally are seen, but a more frequently occurring discrepancy is ± factor of two. At times, differences as high as a factor of five to ten are found. Readers are cautioned that, depending on the model applied, the predictions of concentrations for a given scenario typically vary by about a factor of two. These differences are similar to those reported by Brighton

TABLE 9-3. Predicted Cloud Centerline Concentrations (at Ground-Level) in vppm (volume parts per million) for All Models, All Scenarios, and All Sensitivity Runs for Downwind Distances of 100 m, 200 m, 500 m, 1600 m, 1500 m, and 2000 m (see Appendices A–G for Complete Descriptions and Analyses)

		SCENARIO 1			SCENARIO 2			SCENARIO 3			SCENARIO 4		SCENARIO 5		SCENARIO 6			SCENARIO 7	
		DEGADS	HGSYSTEM	SLAB	DEGADS	HGSYSTEM	SLAB	ALOHA	DEGADS	HGSYSTEM	DEGADS	SLAB	INPUFF	ISC	ALOHA	HGSYSTEM	SLAB	DEGADS	HGSYSTEM
BASE RUN	100 m	887000	6619	18170	173550	25515	17370	655	2585	1916	27180	22620	0.051	9.4	9120	1311	4486	9510	11924
	200 m	33284	3270	8040	10389	8687	7995	232	935	599	14935	11700	0.88	4.1	2980	583	1690	2112	883
	500 m	1399	1550	2735	987	2562	2698	48	160	109	1991	4384	1.8	0.8	574	156	362	363	395
	1000 m	296	784	1212	259	1102	1181	15	40	29	451	1960	0.95	0.22	158	51	104	96	229
	1500 m	136	390	745	129	680	727	7.8	18	13	211	1207	0.5	0.1	74	27	50	44	152
	2000 m	81	266	532	80	483	515	6	11	7.5	128	857	0.29	0.063	43	17	31	25	108
SENS. RUN 1	100 m	see note	13627	45000	2733	1926	1319	3210	6832	8637	12580	19040	1.41E-12	7.6	19100	4949	7345	9708	28543
	200 m	see note	6339	21900	1233	738	570	905	1437	3405	5217	10070	6.24E-05	5.9	6700	2960	4750	2122	8152
	500 m	52274	2670	7907	238	158	145	184	282	1092	2431	3956	0.061	3.2	1570	1857	2621	365	1837
	1000 m	2459	1490	3652	64	44	46	59	94	403	650	1827	0.39	1.4	528	997	1518	97	438
	1500 m	800	1424	2180	30	21	23	31	52	276	287	1147	0.52	0.79	282	637	1018	44	257
	2000 m	405	872	1547	17	12	14	20	34	188	166	820	0.53	0.64	182	450	747	25	164
SENS. RUN 2	100 m	7423	8289	6042	1416	3385	2766	673	6347	5381	11290	16700	4.77E-06	2.3	22600	5898	5995	68540	83886
	200 m	2074	4170	1990	403	1380	1246	247	2380	1976	3604	6673	4.60E-02	2.6	8500	3306	3090	6527	9116
	500 m	441	601	430	94	422	433	52	611	416	1628	2701	0.83	0.67	2110	514	1086	789	587
	1000 m	147	144	129	34	165	187	16	180	114	757	1336	0.73	0.19	607	120	392	188	365
	1500 m	79	63	62	19	93	111	8.4	78	52	366	882	0.44	0.092	287	53	200	81	258
	2000 m	51	36	38	13	61	77	6.4	43	30.1	210	649	0.26	0.055	167	30	124	45	191
SENS. RUN 3	100 m	52098	7939	9735	11985	8737	8363	655	2612	2340	28165	22620	2.16E-05	0.31	9120	6346	5022	3666	4488
	200 m	2514	4690	4660	1735	3068	4145	232	936	886	15223	11800	0.031	1.5	2980	2776	2200	920	709
	500 m	299	1020	1659	296	1003	1535	48	236	194	1971	4422	0.75	0.6	674	725	590	190	352
	1000 m	88	370	770	101	482	715	15	84	55	449	1997	0.71	0.18	158	233	196	57	172
	1500 m	47	210	485	56	317	452	7.8	42	26	210	1237	0.44	0.09	74	124	99	27	110
	2000 m	31	130	352	38	233	327	5	23	14.8	127	885	0.26	0.054	43	79	62	16	75

Note: The receptor location is still within the "vapor blanket" predicted by the DEGADIS model (see Appendix A).

et al. (1994) in their comparisons of the predictions of eight models for some standard scenarios.

The comparisons of the predictions of these same models with observations from field studies reported in Section 8 reveal that there is also typically a factor of two scatter or error between model predictions and observations for a given field experiment (Hanna et al., 1993). All of the models that were applied to Scenarios 1 through 7 showed a similar level of skill (as measured by mean bias and scatter) when compared with the field data in the Hanna et al. (1993) study.

Because of the random or stochastic aspect of the atmosphere, this amount of error or uncertainty is expected in any given test case or application.

APPENDIX A

Scenario 1: Release of Pressurized Liquid Chlorine

This scenario involves the release of pressurized liquid chlorine resulting from the failure of a horizontally oriented pipe near the base of vertical cylindrical tank of height 5 m and of diameter 4 m (see Figure A-1). The pipe diameter is 0.0381 m (1.5 in) and the pipe length is 3.81 m (12.5 ft). It is assumed that the tank sits on the ground and that the pipe is 1.5 m above the ground. There is 4.04 m of liquid in the tank. The chlorine is stored at ambient temperature (298 K) and is pressurized to 8 atm (i.e., saturation pressure). The ambient atmosphere is assumed to be stable (class F) with a wind speed of 2 m s^{-1} at a height of 10 m. Table 9-2 listed the properties of chlorine, the source configuration, the meteorological data, and the site conditions for this scenario, as well as for the other six scenarios. The information specific to this scenario is repeated below in Table A-1. Explanations of derived parameters and the rationale for the sensitivity runs are provided in later paragraphs in this appendix.

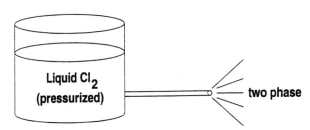

Figure A-1. Schematic diagram of Scenario 1, a pressurized horizontal liquid release of chlorine.

TABLE A-1

Input Data Archive for Scenario 1, a Pressurized Horizontal Liquid Release of Chlorine (Data Taken from Table 9-2, which covers All Seven Scenarios)

CHLORINE PROPERTIES	
molecular weight	$M = 70.91$ kg kmole^{-1}
normal boiling point	$T_b = 239.1$ K
latent heat of evaporation	$H_{vap} = 287,840$ J kg^{-1}
specific heat at constant pressure for vapor	$c_{pv} = 498$ J kg^{-1} K^{-1}
specific heat at constant pressure for liquid	$c_{pl} = 926$ J kg^{-1} K^{-1}
density of liquid	$\rho_l = 1574$ kg m^{-3}
molecular diffusivity	$D_m = 1.22 \times 10^{-5}$ m^2 s^{-1}
kinematic viscosity	$v = 4.48 \times 10^{-6}$ m^2 s^{-1}
SOURCE CONFIGURATION	
storage temperature	$T_s = 298$ K
release temperature	$T_o = 239.1$ K (normal boiling point)
storage pressure	$p_s = 8$ atm
pipe diameter	$D_p = 0.0381$ m
pipe length	$L_p = 3.81$ m
source height	$h_s = 1.5$ m
mass in tank	Mass = 80,000 kg
tank diameter	$D_t = 4$ m
tank height	$h_t = 5$ m
fluid level	$h_t = 4.04$ m
mass flash fraction	$f = 0.189$
two-phase density	$\rho_m = 18.9$ kg m^{-3}
source strength	$M = 11.04$ kg s^{-1}
METEOROLOGICAL CONDITIONS	
temperature	$T_a = 298$ K at height of 10 m
wind speed	$u = 2$ m s^{-1} at height of 10 m
stability class	F
pressure	$p_a = 1$ atm
relative humidity	RH = 80%
Monin-Obukhov length	$L = 12.2$ m
friction velocity	$u* = 0.0808$ m s^{-1}
surface roughness	$z_0 = 0.03$ m (uncut grass)
SENSITIVITY RUNS	
RUN 1	Double source pipe diameter to 0.0762 m, thus multiplying source strength by a factor of 4
RUN 2	Change stability class to D (neutral) and increase wind speed to 5 ms^{-1}
RUN 3	Increase roughness length by one order of magnitude to $z_0 = 0.3$ m
CONCENTRATIONS OF INTEREST	
averaging time $\quad T_a = 600$ sec	IDLH $\quad C = 30$ ppm

The Fauske and Epstein (1988) equation (4-16) can be used to calculate the mass emission rate of the two-phase jet, Q_A, assuming saturated liquid conditions:

$$Q_A = \frac{FH_{vap}}{v_{lg}\sqrt{T_s c_{pl}}} \tag{A-1}$$

where

Q_A = mass flux or emission rate per unit area (kg s^{-1} m^{-2})

F = frictional loss factor

H_{vap} = latent heat of evaporation (J kg^{-1})

v_{lg} = difference in specific volume between gas and liquid (m^3 kg^{-1})($= v_g - v_l = \rho_g-1 - \rho_l-1$)

T_s = storage temperature of the liquid in the tank (K)

c_{pl} = specific heat at constant pressure for liquid (J kg^{-1} K^{-1})

For a pipe length-to-diameter ratio of 100, Fauske and Epstein (1988) suggest a frictional loss factor, F, of 0.75. It is known that $v_g = 4.307 \cdot 10^{-2}$ m^3 kg^{-1} at temperature 298 K and pressure 8 atm, and that $v_l = 6.353 \cdot 10^{-4}$ m^3 kg^{-1}. Equation (A-1) can then be used to calculate that $Q_A = 9685$ kg s^{-1} m^{-2}, which can be multiplied by the pipe cross-sectional area of $1.14 \cdot 10^{-3}$ m^2 to give an emission rate, Q, of 11.04 kg s^{-1} for the two-phase aerosol.

Since the liquid chlorine in the tank is pressurized, some will flash upon release, with the mass fraction, f, that flashes to vapor given by equation (4-14):

$$f = \frac{Q_f}{Q_l} = \frac{c_{pl}(T_s - T_b)}{H_{vap}} \tag{A-2}$$

where

Q_f = mass emission rate of liquid that flashes (kg s^{-1})

Q_l = total liquid mass emission rate (kg s^{-1})

c_{pl} = specific heat at constant pressure for liquid (J kg^{-1} K^{-1})

T_s = storage temperature of the liquid in the tank (K)

T_b = normal boiling point of liquid (K)

H_{vap} = latent heat of evaporation (J kg^{-1})

Table A-1 shows that $c_{pl} = 926$ J kg^{-1} K^{-1}, $T_b = 239.1$ K, and $H_{vap} = 287840$ J kg^{-1}. Consequently, the fraction $f = 0.189$ for this set of conditions.

As a conservative approach, it is assumed that all unflashed liquid is entrained into the gas jet as an aerosol. It is further assumed that the initial cloud temperature is held at the boiling point. The density, ρ_m, of the initial mixture of gas and aerosols can then be calculated from a relation equivalent to equation (5-16):

$$\rho_m = \frac{1}{[(1-f)/\rho_l] + (f/\rho_g)} \qquad \text{(A-3)}$$

where

ρ_l = density of liquid chlorine (1574 kg m^{-3})

ρ_g = density of gaseous chlorine at the boiling point temperature (3.617 kg m^{-3})

Equation (A-3) yields an estimate of the initial mixture density, ρ_m, of 18.9 kg m^{-3}, or about 16 times heavier than air.

Before proceeding with the application of a dense gas dispersion model, it is advisable to calculate the source Richardson number, Ri_0, to see if it exceeds the critical value of 50 (see Section 5.1). Since this scenario involves a continuous elevated jet release, equation (5-3a) should be used:

$$Ri_0 = \frac{g(\rho_{po} - \rho_a)V_{co}}{\rho_a u_*^3 h_s} \qquad \text{(A-4)}$$

The following variables were calculated above, were prescribed as input in Table A-1, or were calculated from known parameters:

ρ_{po} = 18.9 kg m^3 (initial pollutant density = ρ_m)

ρ_a = 1.2 kg m^{-3}

g = 9.8 m s^{-2}

h_s = 1.5 m

u_* = 0.081 m s^{-1}

V_{co} = $Q/(\pi\rho_{po})$ = 0.186 m^3 s^{-1}

By substituting these variables into equation (A-4), it is found that Ri_0 $\approx$ 34,500. Because Ri_0 exceeds the critical value, $Ri_0 = 50$, by several orders of magnitude, it is concluded that it is appropriate to apply dense gas models to this scenario. Otherwise, if $Ri_0 < 50$, a passive gas model such as ISC3 (EPA, 1995) could be applied.

The DEGADIS, HGSYSTEM and SLAB dense gas dispersion models have been applied to this scenario, given the source conditions calculated above. The characteristics of these models and some results of their evaluation with field data were described in Chapters 7 and 8. Both HGSYSTEM and SLAB are able to explicitly treat horizontal jets and aerosol releases. However, in order to simulate aerosol releases with DEGADIS, the user must provide the pollutant mole fraction and the pollutant mass concentration (kg pollutant/m^3 mixture) as a function of the mixture density (kg mixture/m^3 mixture), based on adiabatic mixing of the pollutant with ambient air (Spicer, 1990). Given this user-specified relation, the DEGADIS model is then run in isothermal mode at the ambient temperature. In reality, of course, as droplets evaporate, the temperature of the cloud will decrease.

The actual diameter, 0.0381 m, of the pipe at the emission source is used to initialize HGSYSTEM since that model can explicitly calculate the flash fraction and the expansion of a jet due to flashing. In order to stimulate two-phase jet releases with SLAB, the users guide advises that a larger source area should be input to the model, representing the area of the jet after the pollutant has flashed and formed an aerosol (Ermak, 1990). This "post-flash" area in SLAB is calculated as:

$$A = \frac{\rho_l A_r}{\rho_m} \tag{A-5}$$

where

ρ_m = density of the liquid droplet–vapor (aerosol) mixture [kg m^{-3}, see equation (A-3)]

ρ_l = density of the liquid (kg m^{-3})

A_r = actual area of the pipe rupture (0.00114 m^2)

Thus, $A = 0.095$ m^2 for SLAB applications

The SLAB and HGSYSTEM models can directly simulate the horizontal jet. However, the jet model in DEGADIS is not applicable to horizontal jet releases (Havens, 1988). Spicer (1990) points out that, if DEGADIS is applied to a horizontal jet release, and the user is not concerned about concentrations close to the source (say, downwind distances less than 100 m) where jet effects are expected to be important, then the release can be approximated as a ground-level area-source release with the given mass emission rate, Q, and an effective initial source radius R_{eff}, given by:

$$R_{eff} = \left(\frac{Q}{\rho_m \pi u_1} \right)^{1/2} \tag{A-6}$$

where ρ_m is the density of the pollutant mixture (kg m^{-3}) and u_1 is the wind speed measured at 1 m (m s^{-1})

The wind speed, u_1, equals 0.79 m, as calculated from the wind profile formulas, giving an effective initial source radius, R_{eff}, of 0.49 m for input to DEGADIS.

The three models were each applied to a base case and three sensitivity cases. The changes to the input conditions, involving the pipe diameter, the stability class/wind speed, and the surface roughness for the sensitivity runs are listed near the end of Table A-1. The first run is called the "base run" for each model. The resulting predicted cloud centerline concentrations at six downwind distances (100 m, 200 m, 500 m, 1000 m, 1500 m, and 2000 m) are presented in Table A-2, as well as the half width of the cloud as defined by the lateral distance from the cloud centerline to the 30 ppm contour. Some of these data are plotted in Figures A-2 through A-5. Figure A-2 presents the variation of cloud centerline concentration with distance for the base run for the three models. Figure A-3 contains the same type of plot but for the base run and the three sensitivity runs for SLAB. Figures A-4 and A-5 are similar to Figures A-2 and A-3, respectively, but concerns cloud half width rather than centerline concentration.

The factor of 50 difference in concentration predictions between DEGADIS and the other two models at downwind distances less than about 250 m in Figure A-2 is caused by the fact that DEGADIS is not directly simulating the horizontal jet, but is instead assuming that the initial cloud is a ground level area source. The DEGADIS-predicted ground-level concentration remains at 100% chlorine (i.e., 10^6 ppm) for the first 125 m downwind because the model is simulating the initial cloud as a "vapor blanket." There is much less entrainment into the ground-based cloud being modeled by DEGADIS than into the jet being modeled by SLAB and HGSYSTEM, causing the differences in concentration predictions. The SLAB and HGSYSTEM predictions are within a factor of about two or three of each other at distances on the figure, whereas the DEGADIS predictions are much higher than the others at distances of about 200 m or less and lower than the others at distances greater than about 500 m. The small jump in HGSYSTEM-predicted concentrations at about $x = 700$ m corresponds to the transition from one module to another within the code.

The sensitivity runs with SLAB in Figure A-3 involve doubling the initial diameter (thus multiplying the mass emission rate by 4.0), changing the stability/wind speed from F/2 to D/5, and increasing the surface roughness, z_0, from 0.03 m to 0.3 m. These changes reflect typical uncertainties in worst-case inputs at industrial facilities. The change in pipe

TABLE A-2
Predicted Cloud Centerline Concentration (ppm) and Cloud Half-Width (Lateral Distance to 30 ppm Contour) as a Function of Downwind Distance for DEGADIS, HGSYSTEM, and SLAB for Four Model Input Conditions (Base Run plus Three Sensitivity Runs) for Scenario 1

		SCENARIO 1					
		Centerline Concentration (ppm)			Lateral Distance to 30 ppm (m)		
		DEGADIS	HGSYSTEM	SLAB	DEGADIS	HGSYSTEM	SLAB
Base Run:	100 m	887000	6619	18170	97	44	253
D_p=0.0381m	200 m	33284	3270	8040	485	113	345
F/2	500 m	1399	1550	2735	776	298	467
z_0=0.03m	1000 m	296	784	1212	925	550	568
	1500 m	136	390	745	982	613	633
	2000 m	81	266	532	996	625	681
Sensitivity Run 1:	100 m	see note[1]	13827	45060	see note[1]	39	363
D_p=0.0762m	200 m	see note[1]	6339	21900	see note[1]	109	540
Q=44.16kg/s	500 m	52274	2570	7907	1073	336	787
	1000 m	2459	1490	3552	1720	663	984
	1500 m	800	1424	2180	1969	795	1110
	2000 m	405	872	1547	2115	981	1204
Sensitivity Run 2:	100 m	7423	8289	6042	93	23	56
D/5	200 m	2074	4170	1990	127	80	76
	500 m	441	601	430	177	137	114
	1000 m	147	144	129	205	165	141
	1500 m	79	63	62	201	157	138
	2000 m	51	35	38	171	90	92
Sensitivity Run 3:	100 m	52088	7939	9735	409	66	245
z_0=0.3m	200 m	2514	4690	4560	599	182	322
	500 m	289	1020	1659	724	352	424
	1000 m	88	370	770	763	362	506
	1500 m	47	210	485	732	376	556
	2000 m	31	130	352	612	398	593

---

[1] The receptor location is still within the "vapor blanket" predicted by the DEGADIS model (see text).

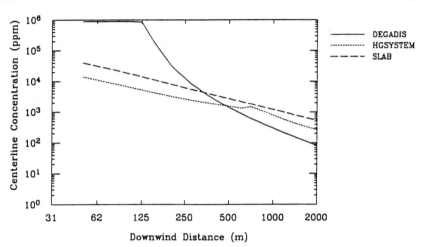

Figure A-2. Predicted cloud centerline concentrations as a function of downwind distance for Scenario 1 base run. Models are DEGADIS, HGSYSTEM and SLAB.

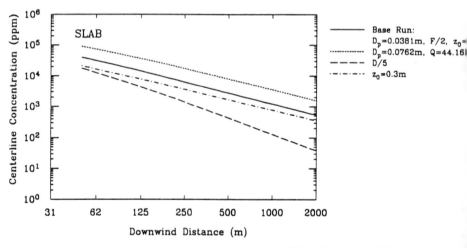

Figure A-3. Predicted cloud centerline concentrations as a function of downwind distance for SLAB for Scenario 1 base run and three sensitivity runs.

diameter is seen to have a large effect on predicted concentrations, in approximate proportion to the increase in mass emission rate. However, the predictions decrease by a factor of about 2 at $x = 50$ m and a factor of about 10 at $x = 2000$ m as the stability changes to class D and the wind speed increases to 5 m s^{-1}. The wind speed change alone should cause a factor of $(5\text{ m s}^{-1})/(2\text{ m s}^{-1}) = 2.5$ decrease due to simple dilution (i.e., concentration is inversely proportional to wind speed). The change from stable to neutral conditions also reduces the concentrations, but the effect is not so great near the source where the cloud is dense. The increase in roughness (from 0.03 m to 0.3 m) causes about a 30% decrease in concentrations, in agreement with previous studies (Hanna et al., 1993) of the effects of changes in roughness on predicted concentrations.

Figure A-4 shows that there are differences in the cloud half-widths (lateral distances to the 30 ppm contour) predicted by the three models,

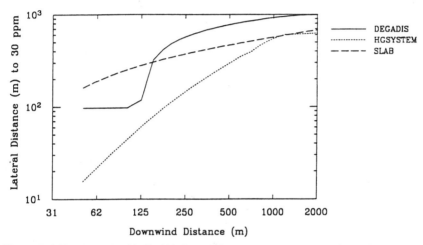

Figure A-4. Predicted cloud half-width (lateral distance to 30 ppm contour) as a function of downwind distance for Scenario 1 base run. Models are DEGADIS, HGSYSTEM, and SLAB.

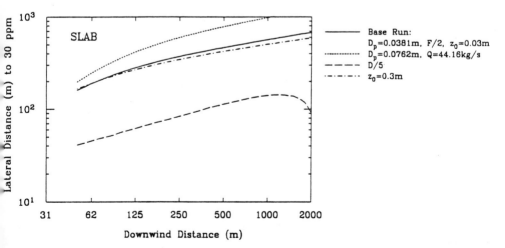

Figure A-5. Predicted cloud half-width (lateral distance to 30 ppm contour) as a function of downwind distance for SLAB for Scenario 1 base run and three sensitivity runs.

especially within about 200 m of the source. At $x = 50$ m, HGSYSTEM predicts a half-width of about 15 m, whereas DEGADIS and SLAB predict half-widths an order of magnitude larger. The DEGADIS-predicted half-width remains constant at 100 m out to a distance of about 100 m because, as mentioned above, the code is simulating the cloud as if it is a 100% "vapor blanket" near the source. By $x = 2000$ m, the model predictions converge to within ±30% of each other. The cloud half-width sensitivity runs with SLAB in Figure A-5 show similar results for the base run, the sensitivity run involving the pipe diameter, and the sensitivity run involving the surface roughness. However, in the case of the sensitivity run involving stability class D/wind speed 5 m s^{-1}, the predicted half-widths are a factor of five less than the others, due to the overall decrease in predicted concentrations everywhere and the associated reduction in spatial coverage of the 30 ppm contour.

As an example of model input/output files, the hard copy HGSYSTEM output files (AEROPLUME for the initial jet and HEGADAS after the jet hits the ground) for the base run of Scenario 1 are listed on the next few pages as Tables A-3 and A-4. These files, as well as similar SLAB and DEGADIS files for all runs are included in the diskette in the pocket inside the back cover.

TABLE A-3 Output File from HGSYSTEM/AEROPLUME for Scenario 1 Base Run

Output from AEROPLUME Version 2.1   Title: Date: 16/08/95 Time: 10:07:03

Reservoir/release orifice conditions		Atmosphere reference conditions		Flash conditions	
reservoir temperature:	24.80 degC	data reference height:	10.00 m	flash temperature:	-34.02 degC
reservoir pressure:	8.00 atm	atmosphere temperature:	24.80 degC	flash pressure:	1.00 atm
orifice diameter:	.04 m	atmosphere pressure:	1.00 atm	flash jet velocity:	76.96 m/s
orifice height:	1.50 m	relative humidity:	80.00 %	flash density:	20.04 kg/m3
orifice pressure:	7.64 atm	ambient wind-speed:	2.00 m/s	flash diameter:	9.55 cm
pollutant mass-flux:	11.04 kg/s	atmosphere density:	1.17 kg/m3	mole fraction liquid:	82.17 %
orifice inclination:	.00 deg	surface roughness:	3.00E-02 m	mole fraction vapour:	17.83 %
orifice velocity:	7.45 m/s	Pasquill/Gifford class:	F	molar mass pollutant:	70.90 kg/kmole
orifice density:	1300.17 kg/m3				
orifice temperature:	21.44 degC				

NB: Reservoir/flash wet pollutant concentration are 100 % by definition.

DX	Z	D	U	PHI	T	H	RHO	CPOL	VPOL	L	DMDT	ENTR	TIME
1.000	1.50	.23	61.77	-.10	-54.90	-196.95	5.214	4.1731	36.376	41.51	13.794	3.507	.014
2.000	1.50	.38	47.77	-.23	-60.90	-145.63	3.245	1.9997	26.001	20.05	17.914	4.600	.033
3.000	1.49	.55	37.54	-.42	-63.60	-108.27	2.559	1.2328	22.196	8.84	22.914	5.317	.057
4.000	1.48	.73	30.34	-.67	-65.05	-82.02	2.231	.8639	20.338	2.43	28.511	5.808	.086
5.000	1.47	.97	25.08	-.98	-39.47	-62.90	1.879	.5976	16.163	2.07	34.704	6.513	.123
6.000	1.45	1.22	21.09	-1.33	-20.52	-48.40	1.673	.4444	12.994	2.06	41.558	7.110	.166
6.999	1.42	1.49	18.05	-1.72	-8.96	-37.34	1.554	.3505	10.717	1.93	48.932	7.572	.218
7.999	1.39	1.76	15.69	-2.15	-1.99	-28.79	1.479	.2879	9.035	1.74	56.716	7.945	.277
8.998	1.35	2.03	13.84	-2.64	.81	-22.11	1.439	.2452	7.775	1.65	64.790	8.156	.345
9.997	1.29	2.32	12.38	-3.19	4.15	-16.78	1.401	.2116	6.796	1.50	73.068	8.375	.422
10.995	1.23	2.61	11.20	-3.65	6.89	-12.53	1.372	.1861	6.031	1.36	81.390	7.613	.507
11.993	1.17	2.97	10.34	-3.11	8.53	-9.58	1.353	.1690	5.511	1.25	88.371	6.559	.600
12.992	1.13	3.44	9.65	-2.15	9.72	-7.30	1.339	.1562	5.114	1.17	94.641	6.030	.700
13.991	1.10	3.98	9.09	-1.10	10.67	-5.42	1.328	.1459	4.791	1.10	100.515	5.793	.807

Interpretation:

DX: horizontal plume centroid displacement (m)
Z: plume-axis (centroid) height (m)
RHO: plume mean density (kg/m3)
CPOL: mean pollutant mass-concentration (kg/m3)
VPOL: volumetric pollutant concentration (vol%)
L: mixture mole-fraction liquid (%)
ENTR: actual entrainment rate (kg/s/m)

H: plume mean mixture enthalpy (kJ/kg)
D: plume (effective) diameter (m)
U: plume mean velocity (m/s)
PHI: plume axis inclination (degrees)
T: mean plume temperature (degC)
DMDT: total plume mass-flux (air+pollutant) (kg/s)
TIME: total elapsed time since discharge (s)

TABLE A-3 (Continued)

DX	Z	D	U	PHI	T	H	RHO	CPOL	VPOL	L	DMDT	ENTR	TIME
14.991	1.09	4.58	8.60	-.17	11.46	-3.78	1.318	.1369	4.510	1.04	106.301	5.784	.920
15.991	1.09	5.21	8.15	.46	12.17	-2.28	1.310	.1289	4.256	.98	117.198	5.741	1.040
16.991	1.10	5.74	7.75	.51	12.77	-.96	1.302	.1219	4.034	.93	117.938	5.742	1.166
17.991	1.11	6.29	7.39	.56	13.30	-.24	1.296	.1157	3.834	.88	123.685	5.754	1.298
18.991	1.12	6.86	7.06	.60	13.78	1.33	1.290	.1100	3.653	.84	129.449	5.775	1.437
19.991	1.13	7.44	6.76	.64	14.21	2.34	1.285	.1049	3.487	.80	135.238	5.803	1.581
20.991	1.14	8.04	6.49	.68	14.60	3.26	1.280	.1001	3.335	.76	141.057	5.836	1.732
21.991	1.15	8.65	6.23	.72	14.95	4.12	1.275	.0958	3.195	.73	146.910	5.871	1.889
22.991	1.17	9.28	6.00	.75	15.27	4.92	1.271	.0918	3.065	.70	152.801	5.910	2.053
23.991	1.18	9.92	5.78	.78	15.57	5.66	1.267	.0881	2.945	.67	158.731	5.950	2.223
24.991	1.19	10.58	5.58	.80	15.84	6.35	1.264	.0847	2.833	.64	164.701	5.990	2.399
25.991	1.21	11.25	5.39	.82	16.10	7.00	1.260	.0815	2.728	.61	170.711	6.031	2.581
26.991	1.22	11.93	5.21	.84	16.33	7.61	1.257	.0785	2.630	.59	176.763	6.072	2.770
27.990	1.24	12.62	5.04	.85	16.55	8.18	1.254	.0757	2.539	.56	182.856	6.113	2.965
28.990	1.25	13.33	4.89	.86	16.75	8.72	1.252	.0731	2.453	.54	188.989	6.153	3.167
29.990	1.27	14.05	4.74	.86	16.95	9.23	1.249	.0707	2.372	.52	195.161	6.192	3.375
30.990	1.28	14.79	4.60	.87	17.12	9.71	1.247	.0684	2.296	.50	201.372	6.230	3.589
31.990	1.30	15.53	4.47	.87	17.29	10.16	1.245	.0662	2.224	.48	207.622	6.268	3.809
32.990	1.31	16.29	4.35	.87	17.45	10.59	1.242	.0641	2.156	.46	213.908	6.304	4.036
33.990	1.33	17.06	4.23	.87	17.60	11.00	1.240	.0622	2.092	.44	220.230	6.340	4.269
34.990	1.34	17.84	4.13	.86	17.73	11.39	1.239	.0603	2.032	.43	226.587	6.374	4.508
35.989	1.36	18.63	4.02	.86	17.87	11.75	1.237	.0586	1.974	.41	232.978	6.408	4.754
36.989	1.37	19.44	3.92	.85	17.99	12.10	1.235	.0570	1.919	.40	239.402	6.440	5.006
37.989	1.39	20.25	3.83	.84	18.11	12.44	1.234	.0554	1.867	.38	245.858	6.472	5.263
38.989	1.40	21.08	3.74	.83	18.22	12.75	1.232	.0539	1.818	.37	252.345	6.502	5.528
39.989	1.42	21.92	3.66	.83	18.33	13.06	1.231	.0525	1.770	.36	258.862	6.531	5.798
40.989	1.43	22.77	3.58	.82	18.43	13.35	1.229	.0511	1.725	.34	265.407	6.560	6.074
41.989	1.45	23.63	3.50	.81	18.52	13.62	1.228	.0498	1.682	.33	271.981	6.588	6.356
42.989	1.46	24.50	3.43	.80	18.61	13.89	1.227	.0486	1.641	.32	278.582	6.614	6.645
43.989	1.47	25.39	3.36	.78	18.70	14.14	1.225	.0474	1.602	.31	285.210	6.640	6.939
44.989	1.49	26.28	3.30	.77	18.78	14.38	1.224	.0463	1.565	.30	291.863	6.665	7.240
45.988	1.50	27.18	3.23	.76	18.86	14.62	1.223	.0452	1.529	.29	298.540	6.689	7.546
46.988	1.51	28.02	3.17	.75	18.94	14.84	1.222	.0442	1.494	.28	305.241	6.713	7.858
47.988	1.53	29.02	3.12	.74	19.01	15.05	1.221	.0432	1.461	.27	311.965	6.735	8.176
48.988	1.54	29.95	3.06	.72	19.08	15.26	1.220	.0423	1.430	.26	318.712	6.757	8.500
49.988	1.55	30.90	3.01	.71	19.15	15.46	1.219	.0414	1.399	.25	325.479	6.778	8.829
54.988	1.61	35.76	2.77	.65	19.44	16.35	1.215	.0373	1.263	.22	359.615	6.872	10.562
59.987	1.67	40.86	2.58	.59	19.68	17.09	1.212	.0339	1.150	.18	394.186	6.951	12.431
64.987	1.71	46.17	2.42	.54	19.89	17.72	1.207	.0311	1.055	.16	429.116	7.016	14.431
69.987	1.76	51.67	2.29	.48	20.06	18.26	1.207	.0287	.973	.13	464.340	7.069	16.556
74.987	1.80	57.36	2.17	.44	20.21	18.72	1.204	.0266	.903	.11	499.800	7.111	18.801

DX	Z	RHO	CPOL	VPOL	L	ENTR	H	D	U	PHI	T	DMDT	TIME
1.84	63.21	1.203	.0248	2.07	.842	.10	535.447	20.34	7.143	.39	19.13	21.161	79.987
1.87	69.21	1.201	.0232	1.99	.789	.08	571.226	20.45	7.165	.35	19.48	23.627	84.987
1.90	75.35	1.200	.0218	1.91	.742	.07	607.097	20.55	7.180	.32	19.80	26.194	89.987
1.92	81.61	1.198	.0206	1.85	.700	.06	643.020	20.64	7.187	.29	20.08	28.857	94.987
1.95	87.99	1.197	.0195	1.79	.662	.04	678.961	20.71	7.188	.26	20.33	31.610	99.986
1.99	101.05	1.195	.0176	1.69	.598	.03	750.782	20.85	7.174	.21	20.76	37.378	109.986
2.02	114.46	1.194	.0160	1.61	.546	.01	822.360	20.95	7.141	.17	21.12	43.454	119.986
2.05	128.14	1.193	.0147	1.54	.502	.00	893.540	21.08	7.095	.14	21.41	49.805	129.986
2.07	141.95	1.191	.0136	1.49	.465	.00	964.114	21.31	7.032	.12	21.67	56.412	139.986
2.09	155.79	1.190	.0127	1.44	.433	.00	1033.902	21.51	6.948	.11	21.88	63.245	149.986
2.11	169.62	1.189	.0119	1.40	.406	.00	1102.784	21.69	6.852	.09	22.06	70.280	159.986
2.12	183.44	1.188	.0112	1.37	.382	.00	1170.659	21.84	6.749	.08	22.23	77.499	169.986
2.13	197.22	1.187	.0106	1.34	.362	.00	1237.467	21.97	6.640	.06	22.37	84.887	179.986
2.14	210.94	1.186	.0100	1.31	.343	.00	1303.167	22.09	6.529	.05	22.49	92.428	189.986
2.15	224.60	1.186	.0096	1.29	.327	.00	1367.739	22.20	6.416	.04	22.60	100.111	199.986
2.16	238.18	1.185	.0091	1.27	.312	.00	1431.184	22.29	6.304	.03	22.70	107.922	209.986
2.16	251.69	1.184	.0088	1.25	.299	.00	1493.526	22.37	6.186	.03	22.79	115.850	219.986
2.17	265.12	1.184	.0084	1.24	.287	.00	1554.738	22.45	6.072	.02	22.87	123.887	229.986
2.17	278.46	1.183	.0081	1.22	.277	.00	1614.833	22.52	5.968	.01	22.95	132.028	239.986
2.17	291.71	1.183	.0078	1.21	.267	.00	1673.846	22.58	5.861	.01	23.02	140.262	249.986
2.17	304.87	1.182	.0075	1.20	.258	.00	1731.799	22.64	5.756	.00	23.08	148.584	259.986
2.17	317.95	1.182	.0073	1.19	.250	.00	1788.714	22.69	5.653	.00	23.13	156.987	269.986
2.17	330.93	1.181	.0071	1.17	.242	.00	1844.616	22.74	5.552	-.00	23.19	165.466	279.986
2.17	343.83	1.181	.0069	1.17	.235	.00	1899.529	22.79	5.454	-.01	23.23	174.017	289.986
2.17	356.64	1.181	.0067	1.16	.229	.00	1953.479	22.83	5.359	-.01	23.28	182.636	299.986
2.17	369.36	1.180	.0065	1.15	.223	.00	2006.492	22.87	5.266	-.01	23.32	191.318	309.986
2.16	381.99	1.180	.0063	1.14	.217	.00	2058.593	22.90	5.176	-.01	23.36	200.060	319.986
2.16	394.54	1.180	.0062	1.13	.212	.00	2109.808	22.94	5.088	-.02	23.40	208.858	329.986
2.16	407.00	1.180	.0060	1.13	.207	.00	2160.161	22.97	5.003	-.02	23.43	217.711	339.986
2.15	419.38	1.180	.0059	1.12	.202	.00	2209.678	23.00	4.920	-.02	23.46	226.615	349.986
2.15	431.69	1.179	.0058	1.11	.198	.00	2258.382	23.03	4.840	-.02	23.49	235.567	359.986
2.15	443.91	1.179	.0056	1.11	.194	.00	2306.298	23.05	4.762	-.02	23.52	244.566	369.986
2.14	456.05	1.179	.0055	1.10	.190	.00	2353.448	23.08	4.686	-.02	23.55	253.609	379.986
2.14	468.12	1.179	.0054	1.10	.186	.00	2399.854	23.10	4.613	-.02	23.57	262.695	389.986
2.14	480.11	1.179	.0053	1.09	.182	.00	2445.539	23.12	4.542	-.02	23.60	271.821	399.986

Interpretation:

DX:	horizontal plume centroid displacement (m)	
Z:	plume-axis (centroid) height (m)	
RHO:	plume mean density (kg/m3)	
CPOL:	mean pollutant mass-concentration (kg/m3)	
VPOL:	volumetric pollutant concentration (vol%)	
L:	mixture mole-fraction liquid (%)	
ENTR:	actual entrainment rate (kg/s/m)	
H:	plume mean mixture enthalpy (kJ/kg)	
D:	plume (effective) diameter (m)	
U:	plume mean velocity (m/s)	
PHI:	plume axis inclination (degrees)	
T:	mean plume temperature (degC)	
DMDT:	total plume mass-flux (air+pollutant) (kg/s)	
TIME:	total elapsed time since discharge (s)	

171

TABLE A-3 (Continued)

DX	Z	D	U	PHI	T	H	RHO	CPOL	VPOL	L	DMDT	ENTR	TIME
409.986	2.13	492.03	1.09	-.02	23.15	23.62	1.179	.0052	.179	.00	2490.524	4.472	280.986
419.986	2.13	503.88	1.08	-.03	23.17	23.64	1.178	.0051	.176	.00	2534.839	4.404	290.188
429.986	2.12	515.66	1.08	-.03	23.19	23.66	1.178	.0050	.173	.00	2578.493	4.337	299.425
439.986	2.12	527.37	1.08	-.03	23.20	23.68	1.178	.0050	.170	.00	2621.505	4.277	308.698
449.986	2.11	539.02	1.07	-.03	23.22	23.70	1.178	.0049	.167	.00	2663.898	4.211	318.003
459.986	2.11	550.60	1.07	-.03	23.24	23.72	1.178	.0048	.165	.00	2705.686	4.151	327.341
469.986	2.10	562.11	1.07	-.03	23.25	23.73	1.178	.0047	.162	.00	2746.885	4.093	336.709
479.986	2.10	573.56	1.06	-.03	23.27	23.75	1.178	.0047	.160	.00	2787.543	4.064	346.108
489.986	2.10	584.96	1.06	-.02	23.28	23.77	1.178	.0046	.158	.00	2828.938	4.195	355.536
499.986	2.09	596.29	1.06	-.01	23.30	23.78	1.178	.0045	.155	.00	2871.682	4.333	364.993
524.986	2.09	624.40	1.05	.01	23.34	23.82	1.177	.0044	.149	.00	2984.614	4.646	388.757
549.986	2.10	652.18	1.04	.02	23.37	23.86	1.177	.0042	.144	.00	3106.104	4.992	412.682
574.986	2.11	679.64	1.04	.03	23.41	23.90	1.177	.0040	.138	.00	3236.199	5.336	436.748
599.986	2.13	706.80	1.03	.04	23.45	23.94	1.177	.0038	.132	.00	3374.984	5.686	460.936
610.811	2.13	718.46	1.03	.05	23.46	23.96	1.176	.0038	.130	.00	3437.795	5.863	471.442

Interpretation:

DX: horizontal plume centroid displacement (m)
Z: plume-axis (centroid) height (m)
RHO: plume mean density (kg/m3)
CPOL: mean pollutant mass-concentration (kg/m3)
VPOL: volumetric pollutant concentration (vol%)
L: mixture mole-fraction liquid (%)
ENTR: actual entrainment rate (kg/s/m)

H: plume mean mixture enthalpy (kJ/kg)
D: plume (effective) diameter (m)
U: plume mean velocity (m/s)
PHI: plume axis inclination (degrees)
T: mean plume temperature (degC)
DMDT: total plume mass-flux (air+pollutant) (kg/s)
TIME: total elapsed time since discharge (s)

TABLE A-4 Output File from HGSYSTEM/HEGADAS for Scenario 1 Base Run

```
HSMAIN                                                                    TIME 10:07
DATE 16/08/95
-------------------------------------------------------------------------------------
                          <<<                          >>>
-------------------------------------------------------------------------------------

=====================================================================================
                        HEGADAS-S INPUT DATA
=====================================================================================

                                             --------> CONTROL data block: control parameters
OUTPUT CODE               ICNT =  0    (no output of cumulative cloud data)
SURFACE-TRANSFER CODE     ISURF = 3    (only heat transfer, no water vapour)
                                             --------> AMBIENT data block: ambient data

AIR TEMP.AT HEIGHT ZAIRTEMP AIRTEMP =  24.800    CELSIUS
REF. HEIGHT FOR AIR TEMP.   ZAIRTEMP =  10.000    M
RELATIVE HUMIDITY           RHPERC  =  80.000    %
WIND VELOCITY AT HEIGHT Z0   U0  =  2.0000    M/S
REFERENCE HEIGHT FOR WIND VEL.  Z0 =  10.000    M
EARTH-S SURFACE TEMPERATURE TGROUND =  24.800    CELSIUS

                                             --------> DISP data block: dispersion data
SURFACE ROUGHNESS PARAMETER    ZR =  3.00000E-02 M
PASQUILL STABILITY CLASS     PQSTAB =  F
AVERAG. TIME FOR CONC.MEAS.  AVTIMC =  600.00    SECONDS
MONIN - OBUKHOV LENGTH        OBUKL =  14.233    M
TYPE OF FORMULA FOR SIGMA_Y  MODSY  =  2         (Briggs formula)
    with parameters:         DELTA  =  4.00000E-02
                             BETA   =  1.00000E-04 M**(-1)
CONST. IN GRAV. SPREADING LAW  CE =  1.1500
CONST. IN GRAV. SPREADING LAW  CD =  5.0000

                                             --------> GASDATA data block: pollutant data
POLLUTANT COMPOSITION:
- basic dry pollutant                    1.0000    (molar fraction)
- extra ideal gas          EXGASPOL =  .00000    (molar fraction)
- water                    WATERPOL =  .00000    (molar fraction)
```

173

TABLE A-4 (Continued)

```
DATA FOR DRY POLLUTANT:
- evaporation rate           GASFLOW =    11.000    KG/S
- specific heat              CPGAS   =    35.300    J/MOLE/CELSIUS
- molecular weight           MWGAS   =    70.900    KG/KMOLE
- heat group in heat flux    HEATGR  =    15.700
DRY POLLUTANT COMPOUNDS:
- CHLORINE        1.00 (molar fraction)
POLLUTANT ENTHALPY           ENTHPOL =   -17720.    JOULE / MOLE OF POLLUTANT
THERMODYNAMIC MODEL          THERMOD =     1         (aerosol model,no HF)

----------------------------------------------------------> CLOUD data block: control of cloud output

NR. OF SOURCE OUTPUT STEPS   NSOURCE =     4
FIXED OUTPUT STEP            DXFIX   =    10.000    M
NUMBER OF FIXED STEPS        NFIX    =     50
INCREASE FACTOR FOR VAR.STEPS XGEOM  =     1.2000
X AT WHICH CALC. IS STOPPED  XEND    =  3000.0     M

CONC. AT WHICH CALC. STOPS   CAMIN   = 1.00000E-08 KG/M3
UPPER CONCENTRATION LIMIT    CUV     = 3.00000E-03 VOL%
LOWER CONCENTRATION LIMIT    CLV     = 1.50000E-03 VOL%
N.B. Value of CU based on APPROXIMATE conversion of CUV
UPPER CONCENTRATION LIMIT    CU      = 8.70517E-05 KG/M3
N.B. Value of CL based on APPROXIMATE conversion of CLV
LOWER CONCENTRATION LIMIT    CL      = 4.35259E-05 KG/M3

-----------------------------------------------------------> TRANSIT data block(s): breakpoint data

DISPERSION CALCULATIONS DOWNWIND OF BREAKPOINT DISTS =   611.00    M ONLY

BREAKPOINT SPECIFIED AT         DISTS =   611.00    M
- half-width of cloud           WS    =   299.00    M
- molar fraction "FRAMOL"       CONCS = 1.83400E-03

WIND PROFILE EXPONENT           ALPHA =    .46621
FRICTION VELOCITY               USTAR = 7.69230E-02 M/S
AIR TEMP. AT GROUND LEVEL       TAP   =   23.878    CELSIUS
```

TABLE A-4 (Continued)

<<<   >>>>

DISTANCE (M)	CONC (% VOL.)	SZ (M)	SY (M)	MIDP (M)	YCU (M)	ZCU (M)	YCL (M)	ZCL (M)	RIB	TMP (DEG.C)	CA (KG/M3)	AEROSOL (%VOL.)
- breakpoint at DISTS =	611.		.000	m								
611.	.183	3.90	.000	299.	299.	10.2	299.	11.4	46.6	22.7	5.36E-03	.0
623.	.179	3.92	16.3	291.	324.	10.2	327.	11.4	45.3	22.8	5.22E-03	.0
637.	.173	3.93	24.3	292.	341.	10.2	345.	11.4	43.9	22.8	5.05E-03	.0
655.	.166	3.96	31.4	295.	358.	10.2	363.	11.4	42.3	22.9	4.85E-03	.0
675.	.158	4.01	38.3	300.	376.	10.3	382.	11.5	40.6	23.0	4.62E-03	.0
700.	.150	4.06	45.3	306.	395.	10.3	403.	11.5	38.8	23.1	4.36E-03	.0
730.	.140	4.15	52.6	313.	417.	10.4	426.	11.6	36.9	23.2	4.07E-03	.0
766.	.129	4.26	60.4	323.	440.	10.5	450.	11.8	34.9	23.3	3.75E-03	.0
809.	.117	4.40	68.7	333.	465.	10.7	477.	12.0	33.0	23.4	3.41E-03	.0
861.	.104	4.60	77.6	345.	492.	10.9	505.	12.3	31.1	23.5	3.04E-03	.0
923.	9.180E-02	4.85	87.3	358.	520.	11.2	536.	12.7	29.5	23.6	2.67E-03	.0
997.	7.938E-02	5.17	97.9	372.	549.	11.6	567.	13.2	28.2	23.7	2.31E-03	.0
1.086E+03	6.764E-02	5.57	109.	386.	580.	12.1	600.	13.9	27.4	23.8	1.97E-03	.0
1.193E+03	5.689E-02	6.11	122.	394.	603.	12.8	626.	14.7	27.5	23.9	1.66E-03	.0
1.321E+03	4.773E-02	6.87	136.	383.	609.	13.8	635.	16.0	29.1	23.9	1.39E-03	.0
1.475E+03	4.015E-02	7.72	150.	371.	614.	14.8	644.	17.4	31.3	24.0	1.17E-03	.0
1.660E+03	3.389E-02	8.64	166.	359.	618.	15.8	653.	18.8	34.4	24.0	9.86E-04	.0
1.882E+03	2.870E-02	9.64	184.	346.	623.	16.8	662.	20.2	38.2	24.0	8.35E-04	.0
2.148E+03	2.440E-02	10.7	203.	333.	626.	17.8	671.	21.6	43.0	24.1	7.10E-04	.0
2.468E+03	2.079E-02	11.9	223.	318.	629.	18.7	681.	23.0	48.8	24.1	6.05E-04	.0
2.851E+03	1.776E-02	13.2	246.	304.	632.	19.5	691.	24.4	55.6	24.1	5.17E-04	.0
3.311E+03	1.519E-02	14.5	271.	288.	633.	20.2	700.	25.8	63.8	24.1	4.42E-04	.0

CPU-seconds for this HEGADAS-S run = 0

APPENDIX B

Scenario 2: Liquid (Cryogenic) Spill of Refrigerated Chlorine

Scenario 2 involves the same chemical (chlorine) and the same tank as Scenario 1. However, the chlorine is now assumed to be refrigerated at 239.1 K rather than stored at an ambient temperature of 298 K, and is now assumed to be stored under ambient pressure rather than at 8 atm. Another difference is that the rupture is on the surface of the tank itself. As a result, the release involves a liquid spill and subsequent evaporation rather than a two-phase jet. Once the liquid chlorine is spilled on the ground, the liquid boils (i.e., the release is cryogenic), since the boiling point temperature (T_b = 239.1 K) for chlorine is lower than the ambient air and ground temperatures (T_a = 298 K). A schematic diagram of Scenario 2 is given in Figure B-1, and the input parameters for the model applications are given in Tables B-1 and 9-2.

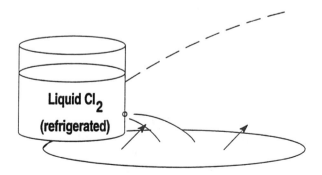

Figure B-1. Schematic diagram of Scenario 2, a refrigerated liquid release of chlorine.

TABLE B-1
Input Data Archive for Scenario 2, a Refrigerated Liquid Release of Chlorine
(Data Taken from Table 9-2, which covers All Seven Scenarios)

CHLORINE PROPERTIES	
molecular weight	$M = 70.91$ kg kmole^{-1}
normal boiling point	$T_b = 239.1$ K
latent heat of evaporation	$H_{vap} = 287{,}840$ J kg^{-1}
specific heat at constant pressure for vapor	$c_{pv} = 498$ J kg^{-1} K^{-1}
specific heat at constant pressure for liquid	$c_{pl} = 926$ J kg^{-1} K^{-1}
density of liquid	$\rho_l = 1574$ kg m^{-3}
molecular diffusivity	$D_m = 1.22 \times 10^{-5}$ m^2 s^{-1}
kinematic viscosity	$v = 4.48 \times 10^{-6}$ m^2 s^{-1}
SOURCE CONFIGURATION	
storage temperature	$T_s = 239.1$ K
release temperature	$T_o = 239.1$ K (normal boiling point)
storage pressure	$p_s = 1$ atm
pipe diameter	$D_p = 0.0381$ m
source height	$h_s = 1.5$ m
mass in tank	Mass = 80,000 kg
tank diameter	$D_t = 4$ m
tank height	$h_t = 5$ m
fluid level	$h_t = 4.04$ m
ground surface	concrete
source strength	$Q = 9.3$ kg s^{-1}
pool radius	$R = 20$ m
METEOROLOGICAL CONDITIONS	
temperature	$T_a = 298$ K at height of 10 m
wind speed	$u = 2$ m s^{-1} at height of 10 m
stability class	F
pressure	$p_a = 1$ atm
relative humidity	RH = 80%
Monin-Obukhov length	$L = 12.2$ m
friction velocity	$u^* = 0.0808$ m s^{-1}
surface roughness	$z_0 = 0.03$ m (uncut grass)
SENSITIVITY RUNS	
RUN 1	Change stability class to D (neutral) and increase wind speed to 10 m s^{-1}
RUN 2	Change radius of liquid pool to 5 m from 20 m
RUN 3	Increase roughness length by one order of magnitude to $z_0 = 0.3$ m
CONCENTRATIONS OF INTEREST	
averaging time $T_a = 600$ sec	IDLH $C = 30$ ppm

The liquid spill rate through the rupture can be calculated using the Bernoulli equation (4-10):

$$Q_1 = c_o A \rho_1 \left[2\left(\frac{p - p_a}{\rho_1}\right) + 2gH \right]^{1/2} \qquad \text{(B-1)}$$

where

Q_1 = liquid mass emission rate (kg s^{-1})

c_o = discharge coefficient (assumed = 1.0 as a conservative estimate)

A = rupture area (m^2)

ρ_1 = liquid density (kg m^{-3})

p = tank pressure (Pa)

p_a = ambient pressure (Pa)

g = acceleration due to gravity (9.8 m s^{-2})

H = height of the liquid above the rupture (m)

From Table 9-2 and Table B-1, $A = 1.14 \cdot 10^{-3}$ m^2, $\rho_1 = 1574$ kg m^{-3}, $p = p_a = 101325$ Pa (1 atm), and $H = 4.04 - 1.50 = 2.54$ m. Therefore, the liquid spill rate, Q_1, initially equals 12.66 kg s^{-1}.

The evaporative emission rate for the cryogenic pool can be calculated using models described in Section 4.2.6. The LPOOL evaporation module of the HGSYSTEM modeling system was used to estimate the time variation of the evaporative emission rate and the size of the pool (Cavanaugh et al., 1994). LPOOL allows the user to specify the discharge coefficient, c_o, for the rupture, whereas models such as ALOHA automatically assumes a discharge coefficient of 0.61. In order to assure that our predictions with LPOOL would be conservative, c_o was set equal to its maximum value, 1.0. It is further assumed that there is no dike and that therefore the liquid can spread unimpeded over the flat concrete surface. The evaporation rates predicted by LPOOL over the first hour after the initial release yielded an average evaporation rate of 9.3 kg s^{-1} and a pool radius of 20 m (see in Tables B-1 and 9-2). For comparisons, the ALOHA "PUDDLE" module was also run giving an average evaporative rate of 6.2 kg s^{-1}. The difference between the LPOOL and ALOHA evaporation is almost entirely explained by the differences in discharge coefficients, c_o (1.0 for LPOOL and 0.61 for ALOHA).

If the predictions of the Bernoulli equation (B-1) were averaged over the first hour of the release, the calculated average mass flow rate through the rupture, Q_1, is 9.77 kg s^{-1}, which is within 5% of the average evaporative emission rate predicted by LPOOL. This near-equality is most likely caused by the fact that, for the special case of an undiked liquid release, the pool radius grows until a near-steady state is reached where the evaporation rate roughly balances the liquid spill rate.

Before proceeding with the application of the dense gas dispersion model, the source Richardson number, Ri_o, should be calculated to see if it exceeds the critical value of about 50 (see Section 5.1). Equation (5-1) should be used for continuous ground-level area sources:

$$Ri_o = \frac{g(\rho_{po} - \rho_a)V_{co}}{\rho_a D_o u_*^3}$$

(B-2)

The following input parameters are prescribed in Table B-1 or can be calculated from known parameters:

ρ_{po} = 3.62 kg m^{-3}

ρ_a = 1.2 kg m^{-3}

g = 9.8 m s^{-2}

u_* = 0.081 m s^{-1}

V_{co} = $Q/(\pi\rho_{po})$ = 0.818 m^3 s^{-1}

By substituting these variables into equation (B-2), it is found that Ri_o = 782, which exceeds the critical Ri_o by over an order of magnitude. Therefore it is appropriate to apply dense gas models to Scenario 2. If Ri_o had been much less than 50, it would be more appropriate to apply a standard passive gas dispersion model such as ISC3 (EPA, 1995).

The DEGADIS, HGSYSTEM and SLAB vapor cloud dispersion models are applied to this scenario, given the source conditions calculated above. All models were initially developed for scenarios involving ground-level area-source releases of dense gases and are therefore directly relevant to this evaporative chlorine source. It is assumed that all models should be run in steady-state mode, since the evaporative emission rate calculated by LPOOL and ALOHA varies by less than a factor of two over a time period of one hour.

The three models were each applied to a base case and to three sensitivity cases. The changes to the input conditions, involving the stability class/wind speed, the pool diameter, and the surface roughness for the sensitivity runs are listed near the end of Table B-1. The first run is called the "base run" for each model. Predicted concentration and cloud half-width variations with downwind distance are listed in Table B-2 and are plotted in Figures B-2 through B-5. This table and these figures are analogous to those used for Scenario 1, which involved the pressurized release of liquid chlorine (see Table A-2 and Figures A-2 through A-5).

TABLE B-2
Predicted Cloud Centerline Concentration (ppm) and Cloud Half-Width (Lateral Distance to 30 ppm Contour) as a Function of Downwind Distance for DEGADIS, HGSYSTEM, and SLAB for Four Model Input Conditions (Base Run plus Three Sensitivity Runs) for Scenario 2

		SCENARIO 2					
		Centerline Concentration (ppm)			Lateral Distance to 30 ppm (m)		
		DEGADIS	HGSYSTEM	SLAB	DEGADIS	HGSYSTEM	SLAB
Base Run:	100 m	173530	25515	17370	181	206	251
F/2	200 m	10389	8687	7995	341	227	315
No dike	500 m	987	2562	2698	484	259	410
z_0=0.03m	1000 m	259	1102	1181	569	292	496
	1500 m	129	680	727	600	316	555
	2000 m	80	483	515	603	335	600
Sensitivity Run 1:	100 m	2733	1926	1319	45	46	44
D/10	200 m	1233	738	570	59	61	54
	500 m	238	158	145	84	86	74
	1000 m	64	44	46	89	74	67
	1500 m	30	21	23	n/a	n/a	n/a
	2000 m	17	12	14	n/a	n/a	n/a
Sensitivity Run 2:	100 m	1415	3385	2766	106	65	89
Diked, D_d=5m	200 m	403	1380	1246	128	73	109
	500 m	94	422	433	146	89	143
	1000 m	34	165	187	110	105	171
	1500 m	19	93	111	n/a	111	182
	2000 m	13	61	77	n/a	108	181
Sensitivity Run 3:	100 m	11985	8737	8363	277	199	244
z_0=0.3m	200 m	1735	3088	4145	357	216	295
	500 m	296	1003	1535	432	239	376
	1000 m	101	482	715	459	262	450
	1500 m	56	317	452	444	279	498
	2000 m	38	233	327	392	293	534

The models are more comparable for Scenario 2, where they are all run in area-source mode, than for Scenario 1, where DEGADIS approximated the horizontal jet by a ground-level area source. Table B-2 and Figure B-2 show that the HGSYSTEM and SLAB concentration predictions are very close (within ±10%) over much of the distance range. However, the DEGADIS concentration predictions are initially larger (due to a calculated

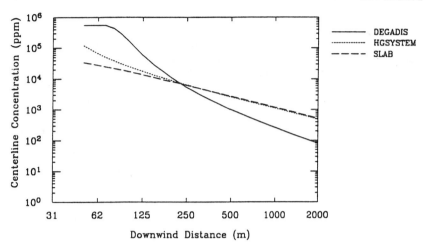

Figure B-2. Predicted cloud centerline concentrations as a function of downwind distance for Scenario 2 base run. Models are DEGADIS, HGSYSTEM and SLAB.

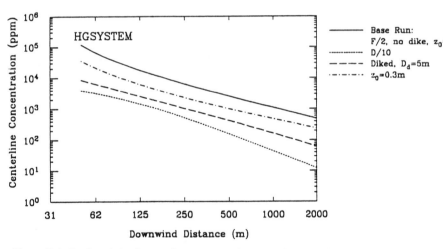

Figure B-3. Predicted cloud centerline concentrations as a function of downwind distance for HGSYSTEM for Scenario 2 base run and three sensitivity runs.

"vapor blanket" over and downwind of the source), but then drop rapidly to values a factor of six less than the other models' predictions at $x = 2000$ m. HGSYSTEM and SLAB suggest a much more gradual decrease of concentration with distance ($C \propto x^{-1}$) than DEGADIS ($C \propto x^{-3}$). Figure B-4 suggests that the predicted cloud half-widths (lateral distances to the 30 ppm contour) are in fair agreement ($\pm$ a factor of two) at distances of greater than about 100 m. At close distances ($x < 100$ m), the DEGADIS

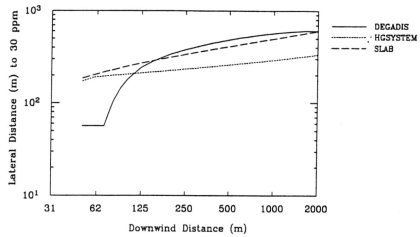

Figure B-4. Predicted cloud half-width (lateral distance to 30 ppm contour) as a function of downwind distance for Scenario 2 base run. Models are DEGADIS, HGSYSTEM, and SLAB.

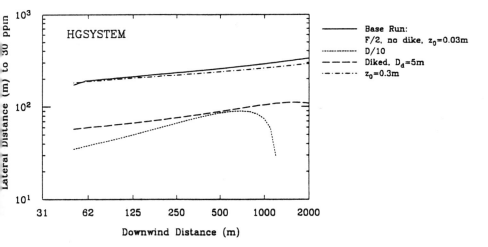

Figure B-5. Predicted cloud half-width (lateral distance to 30 ppm contour) as a function of downwind distance for HGSYSTEM for Scenario 2 base run and three sensitivity runs.

predicted half-widths are less due to the predicted persistence of the dense "vapor blanket."

The results of the sensitivity runs are available for all models (see the results on the floppy disk), but are plotted here only for HGSYSTEM. Concentrations are plotted in Figure B-3 and cloud half-widths are plotted in Figure B-5. It is seen that, when wind speed is increased from 2 m s^{-1} to 10 m s^{-1} and stability is changed from class F (stable) to class D (neutral), predicted concentrations decrease by a factor of 10 near the source and a factor of 30 or 40 at distances of 1000 m or 2000 m. This change is caused by a combination of increased dilution due to the increase in wind speed and increased dispersion due to the change in stability. In the sensitivity run where the liquid pool radius is reduced from 20 m to 5 m, the pool area and evaporation rate decrease by a factor of $(20/5)^2 = 16$. It is seen that there is a resulting factor of about eight decrease in predicted concentration. In the sensitivity run where the surface roughness increases by an order of magnitude (from 0.03 m to 0.3 m), predicted concentrations decrease by about a factor of two or three, in agreement with the findings of Hanna et al. (1993).

The sensitivity of predicted cloud half-widths to variations in input parameters is shown in Figure B-5. The sensitivity run involving a change in surface roughness did not affect the predicted cloud half-width by more than about 10% (lateral distance to the 30 ppm contour). However, the sensitivity runs involving changes in stability class/wind speed and in pool area caused a factor of about three to five decrease in predicted half-width. Note that the predicted cloud half-width drops rapidly to zero beyond $x = 1000$ m for the D/10 sensitivity run, since cloud centerline concentrations are less than 30 ppm beyond that distance.

As an example of model input/output files, the hard copy SLAB output file for the base run of Scenario 2 are listed in the next few pages as Table B-3. These files as well as similar HGSYSTEM and DEGADIS files for all runs are provided in the diskette in the pocket inside the back cover.

TABLE B-3 Output File for SLAB for Scenario 2 Base Run

problem input

idspl	=	1	rhosl	=	1574.00	hs	=	0.00	za	=	10.00

idspl = 1	rhosl = 1574.00	hs = 0.00	za = 10.00
ncalc = 1	spb = -1.00	tav = 600.00	ua = 2.00
wms = 0.070910	spc = 0.00	xffm = 2500.00	ta = 298.00
cps = 498.00	ts = 239.10	zp(1) = 0.00	rh = 80.00
tbp = 239.10	qs = 9.30	zp(2) = 0.00	ppm = 30.00
cmed0 = 0.00	as = 1257.00	zp(3) = 0.00	stab = 0.00
dhe = 287800.	tsd = 3600.	zp(4) = 0.00	ala = 0.0817
cpsl = 926.00	qtis = 33480.00	z0 = 0.030000	

release gas properties

molecular weight of source gas (kg)	- wms	=	7.0910E-02
vapor heat capacity, const. p. (j/kg-k)	- cps	=	4.9800E+02
temperature of source gas (k)	- ts	=	2.3910E+02
density of source gas (kg/m3)	- rhos	=	3.6143E+00
boiling point temperature	- tbp	=	2.3910E+02
liquid mass fraction	- cmed0=		0.0000E+00
liquid heat capacity (j/kg-k)	- cpsl	=	9.2600E+02
heat of vaporization (j/kg)	- dhe	=	2.8780E+05
liquid source density (kg/m3)	- rhosl	=	1.5740E+03
saturation pressure constant	- spa	=	1.0268E+01
saturation pressure constant (k)	- spb	=	2.4546E+03
saturation pressure constant (k)	- spc	=	0.0000E+00
concentration of interest (ppm)	- ppm	=	3.0000E+01

spill characteristics

spill type	- idspl	=	1
mass source rate (kg/s)	- qs	=	9.3000E+00
continuous source duration (s)	- tsd	=	3.6000E+03
continuous source mass (kg)	- qtcs	=	3.3480E+04
instantaneous source mass (kg)	- qtis	=	0.0000E+00
source area (m2)	- as	=	1.2570E+03
vertical vapor velocity (m/s)	- ws	=	2.0471E-03
source half width (m)	- bs	=	1.7727E+01
source height (m)	- hs	=	0.0000E+00
horizontal vapor velocity (m/s)	- us	=	0.0000E+00

field parameters

concentration averaging time (s)	- tav	=	6.0000E+02
mixing layer height (m)	- hmx	=	2.2532E+02
maximum downwind distrace (m)	- xffm	=	2.5000E+03
concentration measurement height (m)	- zp(1)	=	0.0000E+00
	- zp(2)	=	0.0000E+00
	- zp(3)	=	0.0000E+00
	- zp(4)	=	0.0000E+00

ambient meteorological properties

molecular weight of ambient air (kg)	- wmae	=	2.8675E-02
heat capacity of ambient air at const p. (j/kg-k)	- cpaa	=	1.0196E+03
density of ambient air (kg/m3)	- rhoa	=	1.1727E+00
ambient measurement height (m)	- za	=	1.0000E+01
ambient atmospheric pressure (pa=n/m2=j/m3)	- pa	=	1.0132E+05
ambient wind speed (m/s)	- ua	=	2.0000E+00
ambient temperature (k)	- ta	=	2.9800E+02
relative humidity (percent)	- rh	=	8.0000E+01
ambient friction velocity (m/s)	- uastr	=	6.0498E-02
atmospheric stability class value	- stab	=	6.2065E+02
inverse monin-obukhov length (1/m)	- ala	=	8.1730E-02
surface roughness height (m)	- z0	=	3.0000E-02

additional parameters

sub-step multiplier	- ncalc	=	1
number of calculational sub-steps	- nssm	=	3
acceleration of gravity (m/s2)	- grav	=	9.8067E+00
gas constant (j/mol- k)	- rr	=	8.3143E+00
von karman constant	- xk	=	4.1000E-01

TABLE B-3 (Continued)

instantaneous spatially averaged cloud parameters

x	zc	h	bb	b	bbx	bx	cv	rho	t	u	ua
-3.65E+01	0.00E+00	0.00E+00	3.65E+01	3.28E+01	3.65E-01	3.65E+01	0.00E+00	1.17E+00	2.98E+02	0.00E+00	0.00E+00
-2.92E+01	0.00E+00	1.47E+00	3.71E+01	3.31E+01	4.10E-01	4.10E+01	5.36E-03	1.18E+00	2.98E+02	5.47E-01	5.83E-01
-2.19E+01	0.00E+00	1.62E+00	3.79E+01	3.36E+01	4.55E-01	4.55E+01	9.58E-03	1.19E+00	2.97E+02	5.46E-01	6.12E-01
-1.46E+01	0.00E+00	1.70E+00	3.95E+01	3.48E+01	5.01E-01	5.01E+01	1.33E-02	1.20E+00	2.97E+02	5.36E-01	6.29E-01
-7.29E+00	0.00E+00	1.74E+00	4.23E+01	3.70E+01	5.46E-01	5.46E+01	1.66E-02	1.21E+00	2.97E+02	5.21E-01	6.36E-01
-8.58E-06	0.00E+00	1.74E+00	4.66E+01	4.04E+01	5.91E-01	5.91E+01	1.95E-02	1.21E+00	2.97E+02	5.04E-01	6.36E-01
7.29E+00	0.00E+00	1.70E+00	5.25E+01	4.52E+01	6.37E-01	6.37E+01	2.19E-02	1.21E+00	2.97E+02	4.88E-01	6.29E-01
1.46E+01	0.00E+00	1.68E+00	5.99E+01	5.12E+01	6.82E-01	6.82E+01	2.37E-02	1.22E+00	2.97E+02	4.68E-01	6.24E-01
2.19E+01	0.00E+00	1.57E+00	6.87E+01	5.84E+01	7.27E-01	7.27E+01	2.50E-02	1.22E+00	2.97E+02	4.73E-01	6.04E-01
2.92E+01	0.00E+00	1.53E+00	7.77E+01	6.57E+01	7.73E-01	7.73E+01	2.54E-02	1.22E+00	2.97E+02	4.77E-01	5.94E-01
3.65E+01	0.00E+00	1.50E+00	8.67E+01	7.30E+01	8.18E-01	8.18E+01	2.53E-02	1.22E+00	2.97E+02	4.83E-01	5.88E-01
3.70E+01	0.00E+00	1.49E+00	8.73E+01	7.35E+01	8.24E-01	8.24E+01	2.50E-02	1.22E+00	2.97E+02	4.86E-01	5.87E-01
3.76E+01	0.00E+00	1.49E+00	8.80E+01	7.41E+01	8.32E-01	8.32E+01	2.48E-02	1.22E+00	2.97E+02	4.88E-01	5.86E-01
3.82E+01	0.00E+00	1.48E+00	8.88E+01	7.47E+01	8.40E-01	8.40E+01	2.45E-02	1.22E+00	2.97E+02	4.90E-01	5.85E-01
3.90E+01	0.00E+00	1.48E+00	8.97E+01	7.54E+01	8.49E-01	8.49E+01	2.42E-02	1.22E+00	2.97E+02	4.93E-01	5.84E-01
3.98E+01	0.00E+00	1.47E+00	9.07E+01	7.63E+01	8.60E-01	8.60E+01	2.39E-02	1.22E+00	2.97E+02	4.95E-01	5.83E-01
4.08E+01	0.00E+00	1.47E+00	9.18E+01	7.72E+01	8.72E-01	8.72E+01	2.35E-02	1.22E+00	2.97E+02	4.99E-01	5.82E-01
4.19E+01	0.00E+00	1.46E+00	9.31E+01	7.83E+01	8.86E-01	8.86E+01	2.31E-02	1.22E+00	2.97E+02	5.03E-01	5.81E-01
4.32E+01	0.00E+00	1.46E+00	9.46E+01	7.94E+01	9.01E-01	9.01E+01	2.25E-02	1.22E+00	2.97E+02	5.08E-01	5.79E-01
4.46E+01	0.00E+00	1.45E+00	9.62E+01	8.08E+01	9.19E-01	9.19E+01	2.20E-02	1.21E+00	2.97E+02	5.13E-01	5.78E-01
4.63E+01	0.00E+00	1.45E+00	9.80E+01	8.23E+01	9.40E-01	9.40E+01	2.13E-02	1.21E+00	2.97E+02	5.18E-01	5.77E-01
4.81E+01	0.00E+00	1.45E+00	1.00E+02	8.39E+01	9.63E-01	9.63E+01	2.06E-02	1.21E+00	2.97E+02	5.24E-01	5.77E-01
5.03E+01	0.00E+00	1.45E+00	1.02E+02	8.57E+01	9.90E-01	9.89E+01	1.98E-02	1.21E+00	2.97E+02	5.30E-01	5.78E-01
5.27E+01	0.00E+00	1.45E+00	1.05E+02	8.78E+01	1.02E+02	1.02E+02	1.90E-02	1.21E+00	2.97E+02	5.37E-01	5.80E-01
5.55E+01	0.00E+00	1.46E+00	1.07E+02	9.00E+01	1.05E+02	1.05E+02	1.81E-02	1.21E+00	2.97E+02	5.44E-01	5.82E-01
5.86E+01	0.00E+00	1.47E+00	1.10E+02	9.24E+01	1.09E+02	1.09E+02	1.71E-02	1.21E+00	2.97E+02	5.52E-01	5.86E-01
6.22E+01	0.00E+00	1.49E+00	1.14E+02	9.51E+01	1.14E+02	1.14E+02	1.61E-02	1.21E+00	2.97E+02	5.61E-01	5.91E-01
6.63E+01	0.00E+00	1.51E+00	1.17E+02	9.79E+01	1.19E+02	1.19E+02	1.51E-02	1.20E+00	2.97E+02	5.70E-01	5.96E-01
7.09E+01	0.00E+00	1.54E+00	1.21E+02	1.01E+02	1.25E+02	1.25E+02	1.40E-02	1.20E+00	2.97E+02	5.80E-01	6.04E-01
7.63E+01	0.00E+00	1.57E+00	1.25E+02	1.04E+02	1.31E+02	1.31E+02	1.30E-02	1.20E+00	2.97E+02	5.91E-01	6.12E-01
8.23E+01	0.00E+00	1.61E+00	1.30E+02	1.08E+02	1.39E+02	1.39E+02	1.19E-02	1.20E+00	2.97E+02	6.03E-01	6.21E-01
8.92E+01	0.00E+00	1.66E+00	1.34E+02	1.12E+02	1.47E+02	1.47E+02	1.09E-02	1.20E+00	2.98E+02	6.16E-01	6.32E-01
9.70E+01	0.00E+00	1.72E+00	1.39E+02	1.16E+02	1.57E+02	1.57E+02	9.89E-03	1.19E+00	2.98E+02	6.29E-01	6.44E-01
1.06E+02	0.00E+00	1.78E+00	1.44E+02	1.20E+02	1.68E+02	1.68E+02	8.94E-03	1.19E+00	2.98E+02	6.44E-01	6.57E-01
1.16E+02	0.00E+00	1.86E+00	1.50E+02	1.24E+02	1.81E+02	1.81E+02	8.04E-03	1.19E+00	2.98E+02	6.61E-01	6.72E-01
1.28E+02	0.00E+00	1.94E+00	1.56E+02	1.29E+02	1.95E+02	1.95E+02			2.98E+02		

x	cm	cmv	cmda	cmw	cmwv	wc	vg	ug	w	v	vx
1.41E+02	0.00E+00	2.03E+00	1.62E+02	1.34E+02	2.12E+02	2.12E+02	7.19E-03	1.19E+00	2.98E+02	6.78E-01	6.88E-01
1.56E+02	0.00E+00	2.14E+00	1.68E+02	1.39E+02	2.30E+02	2.30E+02	6.41E-03	1.18E+00	2.98E+02	6.97E-01	7.06E-01
1.73E+02	0.00E+00	2.25E+00	1.75E+02	1.44E+02	2.52E+02	2.52E+02	5.66E-03	1.18E+00	2.98E+02	7.16E-01	7.25E-01
1.93E+02	0.00E+00	2.38E+00	1.82E+02	1.50E+02	2.76E+02	2.76E+02	5.01E-03	1.18E+00	2.98E+02	7.38E-01	7.45E-01
2.15E+02	0.00E+00	2.53E+00	1.89E+02	1.55E+02	3.04E+02	3.04E+02	4.41E-03	1.18E+00	2.98E+02	7.60E-01	7.67E-01
2.40E+02	0.00E+00	2.69E+00	1.97E+02	1.61E+02	3.35E+02	3.35E+02	3.86E-03	1.18E+00	2.98E+02	7.84E-01	7.90E-01
2.69E+02	0.00E+00	2.86E+00	2.05E+02	1.68E+02	3.71E+02	3.71E+02	3.38E-03	1.18E+00	2.98E+02	8.09E-01	8.15E-01
3.02E+02	0.00E+00	3.06E+00	2.13E+02	1.74E+02	4.12E+02	4.12E+02	2.94E-03	1.18E+00	2.98E+02	8.35E-01	8.41E-01
3.40E+02	0.00E+00	3.27E+00	2.22E+02	1.81E+02	4.59E+02	4.59E+02	2.56E-03	1.18E+00	2.98E+02	8.63E-01	8.68E-01
3.82E+02	0.00E+00	3.50E+00	2.32E+02	1.88E+02	5.12E+02	5.12E+02	2.22E-03	1.18E+00	2.98E+02	8.92E-01	8.96E-01
4.31E+02	0.00E+00	3.75E+00	2.42E+02	1.95E+02	5.72E+02	5.72E+02	1.92E-03	1.18E+00	2.98E+02	9.22E-01	9.26E-01
4.86E+02	0.00E+00	4.02E+00	2.52E+02	2.03E+02	6.41E+02	6.41E+02	1.66E-03	1.18E+00	2.98E+02	9.53E-01	9.56E-01
5.49E+02	0.00E+00	4.32E+00	2.63E+02	2.11E+02	7.19E+02	7.19E+02	1.43E-03	1.18E+00	2.98E+02	9.85E-01	9.88E-01
6.21E+02	0.00E+00	4.64E+00	2.75E+02	2.20E+02	8.09E+02	8.09E+02	1.24E-03	1.17E+00	2.98E+02	1.02E+00	1.02E+00
7.03E+02	0.00E+00	4.99E+00	2.87E+02	2.29E+02	9.11E+02	9.11E+02	1.06E-03	1.17E+00	2.98E+02	1.05E+00	1.06E+00
7.96E+02	0.00E+00	5.37E+00	3.00E+02	2.39E+02	1.03E+03	1.03E+03	9.16E-04	1.17E+00	2.98E+02	1.09E+00	1.09E+00
9.03E+02	0.00E+00	5.77E+00	3.14E+02	2.49E+02	1.16E+03	1.16E+03	7.88E-04	1.17E+00	2.98E+02	1.12E+00	1.13E+00
1.02E+03	0.00E+00	6.21E+00	3.29E+02	2.59E+02	1.31E+03	1.31E+03	6.77E-04	1.17E+00	2.98E+02	1.16E+00	1.16E+00
1.16E+03	0.00E+00	6.68E+00	3.45E+02	2.70E+02	1.48E+03	1.48E+03	5.82E-04	1.17E+00	2.98E+02	1.20E+00	1.20E+00
1.32E+03	0.00E+00	7.18E+00	3.61E+02	2.82E+02	1.68E+03	1.68E+03	5.00E-04	1.17E+00	2.98E+02	1.24E+00	1.24E+00
1.50E+03	0.00E+00	7.73E+00	3.79E+02	2.95E+02	1.90E+03	1.90E+03	4.29E-04	1.17E+00	2.98E+02	1.28E+00	1.28E+00
1.70E+03	0.00E+00	8.31E+00	3.99E+02	3.08E+02	2.15E+03	2.15E+03	3.68E-04	1.17E+00	2.98E+02	1.32E+00	1.32E+00
1.93E+03	0.00E+00	8.94E+00	4.19E+02	3.22E+02	2.44E+03	2.44E+03	3.15E-04	1.17E+00	2.98E+02	1.36E+00	1.36E+00
2.20E+03	0.00E+00	9.89E+00	4.41E+02	3.37E+02	2.55E+03	2.55E+03	2.59E-04	1.17E+00	2.98E+02	1.41E+00	1.42E+00
2.53E+03	0.00E+00	1.09E+01	4.65E+02	3.53E+02	2.68E+03	2.68E+03	2.12E-04	1.17E+00	2.98E+02	1.47E+00	1.47E+00

x	cm	cmv	cmda	cmw	cmwv	wc	vg	ug	w	v	vx
-3.65E+01	0.00E+00	0.00E+00	9.84E-01	1.64E-02	1.64E-02	0.00E+00	0.00E+00	0.00E+00	9.47E-01	3.88E-02	0.00E+00
-2.92E+01	0.00E+00	1.31E-01	9.71E-01	1.64E-02	1.62E-02	0.00E+00	1.70E-02	0.00E+00	8.70E-03	9.76E-03	1.40E-01
-2.19E+01	0.00E+00	2.34E-02	9.61E-01	1.60E-02	1.60E-02	0.00E+00	6.41E-02	0.00E+00	5.34E-03	1.12E-02	1.46E-01
-1.46E+01	0.00E+00	3.23E-02	9.52E-01	1.59E-02	1.59E-02	0.00E+00	1.37E-01	0.00E+00	4.06E-03	1.27E-02	1.49E-01
-7.29E+00	0.00E+00	4.02E-02	9.44E-01	1.56E-02	1.57E-02	0.00E+00	2.29E-01	0.00E+00	3.60E-03	1.39E-02	1.51E-01
-8.58E-06	0.00E+00	4.70E-02	9.37E-01	1.56E-02	1.56E-02	0.00E+00	3.29E-01	0.00E+00	3.61E-03	1.48E-02	1.51E-01
7.29E+00	0.00E+00	5.26E-02	9.32E-01	1.55E-02	1.55E-02	0.00E+00	4.21E-01	0.00E+00	3.94E-03	1.50E-02	1.49E-01
1.46E+01	0.00E+00	5.67E-02	9.28E-01	1.54E-02	1.55E-02	0.00E+00	4.96E-01	0.00E+00	4.27E-03	1.58E-02	1.48E-01
2.19E+01	0.00E+00	5.95E-02	9.25E-01	1.54E-02	1.54E-02	0.00E+00	5.51E-01	0.00E+00	5.10E-03	1.37E-02	1.44E-01
2.92E+01	0.00E+00	6.06E-02	9.24E-01	1.54E-02	1.54E-02	0.00E+00	5.69E-01	0.00E+00	5.53E-03	1.25E-02	1.42E-01
3.65E+01	0.00E+00	6.06E-02	9.24E-01	1.54E-02	1.54E-02	0.00E+00	5.69E-01	0.00E+00	5.75E-03	1.15E-02	1.41E-01
3.70E+01	0.00E+00	6.02E-02	9.24E-01	1.54E-02	1.54E-02	0.00E+00	5.68E-01	0.00E+00	5.78E-03	1.14E-02	1.41E-01
3.76E+01	0.00E+00	5.97E-02	9.25E-01	1.54E-02	1.54E-02	0.00E+00	5.68E-01	0.00E+00	5.81E-03	1.12E-02	1.41E-01
3.82E+01	0.00E+00	5.92E-02	9.25E-01	1.54E-02	1.54E-02	0.00E+00	5.67E-01	0.00E+00	5.85E-03	1.10E-02	1.41E-01
3.90E+01	0.00E+00	5.86E-02	9.26E-01	1.54E-02	1.54E-02	0.00E+00	5.66E-01	0.00E+00	5.91E-03	1.08E-02	1.41E-01

TABLE B-3 (Continued)

3.98E-01	5.79E-02	5.79E-02	9.27E-01	1.54E-02	1.54E-02	0.00E+00	5.65E-01	0.00E+00	5.98E-03	1.05E-02	1.40E-01
4.08E-01	5.71E-02	5.71E-02	9.27E-01	1.55E-02	1.55E-02	0.00E+00	5.63E-01	0.00E+00	6.07E-03	1.03E-02	1.40E-01
4.19E-01	5.62E-02	5.62E-02	9.28E-01	1.55E-02	1.55E-02	0.00E+00	5.60E-01	0.00E+00	6.19E-03	9.97E-03	1.40E-01
4.32E-01	5.51E-02	5.51E-02	9.29E-01	1.55E-02	1.55E-02	0.00E+00	5.57E-01	0.00E+00	6.33E-03	9.61E-03	1.40E-01
4.46E-01	5.40E-02	5.40E-02	9.31E-01	1.55E-02	1.55E-02	0.00E+00	5.52E-01	0.00E+00	6.49E-03	9.25E-03	1.39E-01
4.63E-01	5.26E-02	5.26E-02	9.32E-01	1.56E-02	1.55E-02	0.00E+00	5.46E-01	0.00E+00	6.63E-03	8.91E-03	1.39E-01
4.81E-01	5.12E-02	5.12E-02	9.33E-01	1.56E-02	1.56E-02	0.00E+00	5.38E-01	0.00E+00	6.77E-03	8.59E-03	1.39E-01
5.03E-01	4.95E-02	4.95E-02	9.35E-01	1.56E-02	1.56E-02	0.00E+00	5.29E-01	0.00E+00	6.88E-03	8.29E-03	1.39E-01
5.27E-01	4.77E-02	4.77E-02	9.37E-01	1.56E-02	1.56E-02	0.00E+00	5.18E-01	0.00E+00	6.97E-03	8.01E-03	1.40E-01
5.55E-01	4.57E-02	4.57E-02	9.39E-01	1.56E-02	1.56E-02	0.00E+00	5.05E-01	0.00E+00	7.04E-03	7.77E-03	1.40E-01
5.86E-01	4.35E-02	4.35E-02	9.41E-01	1.57E-02	1.57E-02	0.00E+00	4.91E-01	0.00E+00	7.08E-03	7.55E-03	1.40E-01
6.22E-01	4.13E-02	4.13E-02	9.43E-01	1.57E-02	1.57E-02	0.00E+00	4.75E-01	0.00E+00	7.08E-03	7.37E-03	1.41E-01
6.63E-01	3.89E-02	3.89E-02	9.45E-01	1.58E-02	1.58E-02	0.00E+00	4.58E-01	0.00E+00	7.07E-03	7.21E-03	1.42E-01
7.09E-01	3.64E-02	3.64E-02	9.48E-01	1.58E-02	1.58E-02	0.00E+00	4.40E-01	0.00E+00	7.03E-03	7.09E-03	1.43E-01
7.63E-01	3.39E-02	3.39E-02	9.50E-01	1.58E-02	1.58E-02	0.00E+00	4.21E-01	0.00E+00	6.96E-03	7.00E-03	1.44E-01
8.23E-01	3.14E-02	3.14E-02	9.53E-01	1.59E-02	1.59E-02	0.00E+00	4.02E-01	0.00E+00	6.88E-03	6.93E-03	1.46E-01
8.92E-01	2.89E-02	2.89E-02	9.55E-01	1.59E-02	1.59E-02	0.00E+00	3.82E-01	0.00E+00	6.79E-03	6.88E-03	1.48E-01
9.70E-01	2.65E-02	2.65E-02	9.58E-01	1.60E-02	1.60E-02	0.00E+00	3.62E-01	0.00E+00	6.70E-03	6.85E-03	1.50E-01
1.06E+02	2.41E-02	2.41E-02	9.60E-01	1.60E-02	1.60E-02	0.00E+00	3.43E-01	0.00E+00	6.61E-03	6.84E-03	1.52E-01
1.16E+02	2.18E-02	2.18E-02	9.62E-01	1.60E-02	1.60E-02	0.00E+00	3.24E-01	0.00E+00	6.51E-03	6.83E-03	1.55E-01
1.28E+02	1.96E-02	1.96E-02	9.64E-01	1.61E-02	1.61E-02	0.00E+00	3.06E-01	0.00E+00	6.42E-03	6.84E-03	1.58E-01
1.41E+02	1.76E-02	1.76E-02	9.66E-01	1.61E-02	1.61E-02	0.00E+00	2.89E-01	0.00E+00	6.33E-03	6.86E-03	1.61E-01
1.56E+02	1.57E-02	1.57E-02	9.68E-01	1.61E-02	1.61E-02	0.00E+00	2.72E-01	0.00E+00	6.25E-03	6.89E-03	1.65E-01
1.73E+02	1.39E-02	1.39E-02	9.70E-01	1.62E-02	1.62E-02	0.00E+00	2.57E-01	0.00E+00	6.16E-03	6.93E-03	1.69E-01
1.93E+02	1.23E-02	1.23E-02	9.72E-01	1.62E-02	1.62E-02	0.00E+00	2.42E-01	0.00E+00	6.08E-03	6.97E-03	1.73E-01
2.15E+02	1.08E-02	1.08E-02	9.73E-01	1.62E-02	1.62E-02	0.00E+00	2.29E-01	0.00E+00	5.99E-03	7.02E-03	1.78E-01
2.40E+02	9.50E-03	9.50E-03	9.74E-01	1.62E-02	1.62E-02	0.00E+00	2.16E-01	0.00E+00	5.91E-03	7.08E-03	1.83E-01
2.69E+02	8.31E-03	8.31E-03	9.75E-01	1.63E-02	1.63E-02	0.00E+00	2.05E-01	0.00E+00	5.83E-03	7.14E-03	1.88E-01
3.02E+02	7.24E-03	7.24E-03	9.76E-01	1.63E-02	1.63E-02	0.00E+00	1.94E-01	0.00E+00	5.74E-03	7.20E-03	1.94E-01
3.40E+02	6.30E-03	6.30E-03	9.77E-01	1.63E-02	1.63E-02	0.00E+00	1.84E-01	0.00E+00	5.65E-03	7.26E-03	2.00E-01
3.82E+02	5.47E-03	5.47E-03	9.78E-01	1.63E-02	1.63E-02	0.00E+00	1.75E-01	0.00E+00	5.56E-03	7.32E-03	2.06E-01
4.31E+02	4.74E-03	4.74E-03	9.79E-01	1.63E-02	1.63E-02	0.00E+00	1.67E-01	0.00E+00	5.46E-03	7.38E-03	2.12E-01
4.86E+02	4.09E-03	4.09E-03	9.80E-01	1.63E-02	1.63E-02	0.00E+00	1.59E-01	0.00E+00	5.37E-03	7.43E-03	2.19E-01
5.49E+02	3.54E-03	3.54E-03	9.80E-01	1.63E-02	1.63E-02	0.00E+00	1.52E-01	0.00E+00	5.27E-03	7.48E-03	2.26E-01
6.21E+02	3.05E-03	3.05E-03	9.81E-01	1.64E-02	1.63E-02	0.00E+00	1.46E-01	0.00E+00	5.16E-03	7.53E-03	2.34E-01
7.03E+02	2.63E-03	2.63E-03	9.81E-01	1.64E-02	1.63E-02	0.00E+00	1.40E-01	0.00E+00	5.06E-03	7.57E-03	2.41E-01
7.96E+02	2.26E-03	2.26E-03	9.81E-01	1.64E-02	1.64E-02	0.00E+00	1.34E-01	0.00E+00	4.95E-03	7.60E-03	2.49E-01
9.03E+02	1.95E-03	1.95E-03	9.82E-01	1.64E-02	1.64E-02	0.00E+00	1.29E-01	0.00E+00	4.85E-03	7.62E-03	2.57E-01
1.02E+03	1.67E-03	1.67E-03	9.82E-01	1.64E-02	1.64E-02	0.00E+00	1.24E-01	0.00E+00	4.74E-03	7.62E-03	2.65E-01
1.16E+03	1.44E-03	1.44E-03	9.82E-01	1.64E-02	1.64E-02	0.00E+00	1.20E-01	0.00E+00	4.63E-03	7.63E-03	2.73E-01

1.32E+03 1.23E-03 9.82E-01 0.00E+00 1.16E-01 1.64E-02 0.00E+00 4.52E-03 7.63E-03 2.81E-01
1.50E+03 1.06E-03 9.83E-01 0.00E+00 1.11E-01 1.64E-02 0.00E+00 4.41E-03 7.61E-03 2.89E-01
1.70E+03 9.09E-04 9.83E-01 0.00E+00 1.08E-01 1.64E-02 0.00E+00 4.30E-03 7.58E-03 2.97E-01
1.93E+03 7.79E-04 9.83E-01 0.00E+00 1.04E-01 1.64E-02 0.00E+00 4.19E-03 7.54E-03 3.06E-01
2.20E+03 6.41E-04 9.83E-01 0.00E+00 9.65E-02 1.64E-02 0.00E+00 4.03E-03 7.57E-03 3.16E-01
2.53E+03 5.25E-04 9.83E-01 0.00E+00 9.00E-02 1.64E-02 0.00E+00 3.88E-03 7.59E-03 3.27E-01

1

time averaged (tav = 600. s) volume concentration: concentration contour parameters

$c(x,y,z,t) = cc(x) \cdot (erf(xa)-erf(xb)) \cdot (erf(ya)-erf(yb)) \cdot (exp(-za \cdot za)+exp(-zb \cdot zb))$

$c(x,y,z,t)$ = concentration (volume fraction) at (x,y,z,t)
x = downwind distance (m)
y = crosswind horizontal distance (m)
z = height (m)
t = time (s)

erf = error functon
xa = (x-xc+bx)/(sr2*betax)
xb = (x-xc-bx)/(sr2*betax)
ya = (y+b)/(sr2*betac)
yb = (y-b)/(sr2*betac)
exp = exponential function
za = (z-zc)/(sr2*sig)
zb = (z+zc)/(sr2*sig)
sr2 = sqrt(2.0)

x	cc(x)	b(x)	betac(x)	zc(x)	sig(x)
-3.65E+01	0.00E+00	3.28E+01	9.18E+00	0.00E+00	0.00E+00
-2.92E+01	1.04E-03	3.31E+01	9.67E+00	0.00E+00	8.50E-01
-2.19E+01	1.87E-03	3.36E+01	1.01E+01	0.00E+00	9.33E-01
-1.46E+01	2.61E-03	3.48E+01	1.08E+01	0.00E+00	9.83E-01
-7.29E+00	3.29E-03	3.70E+01	1.19E+01	0.00E+00	1.01E+00
-8.58E-06	3.90E-03	4.04E+01	1.35E+01	0.00E+00	1.00E+00
7.29E+00	4.41E-03	4.52E+01	1.55E+01	0.00E+00	9.83E-01
1.46E+01	4.81E-03	5.12E+01	1.81E+01	0.00E+00	9.68E-01
2.19E+01	5.08E-03	5.84E+01	2.11E+01	0.00E+00	9.09E-01
2.92E+01	5.21E-03	6.57E+01	2.41E+01	0.00E+00	8.81E-01
3.65E+01	5.23E-03	7.30E+01	2.71E+01	0.00E+00	8.65E-01
3.70E+01	5.19E-03	7.35E+01	2.73E+01	0.00E+00	8.62E-01
3.76E+01	5.15E-03	7.41E+01	2.76E+01	0.00E+00	8.60E-01
3.82E+01	5.10E-03	7.47E+01	2.78E+01	0.00E+00	8.57E-01

t	xc(t)	bx(t)	betax(t)
3.48E+01	0.00E+00	3.65E+01	2.98E-01
4.29E+01	3.65E+00	4.10E+01	3.35E-01
5.10E+01	7.29E+00	4.55E+01	3.72E-01
6.09E+01	1.09E+01	5.01E+01	4.09E-01
7.08E+01	1.46E+01	5.46E+01	4.46E-01
8.27E+01	1.82E+01	5.91E+01	4.83E-01
9.46E+01	2.19E+01	6.37E+01	5.20E-01
1.08E+02	2.55E+01	6.82E+01	5.57E-01
1.21E+02	2.92E+01	7.27E+01	5.94E-01
1.35E+02	3.28E+01	7.73E+01	6.31E-01
1.50E+02	3.65E+01	8.18E+01	6.68E-01
1.52E+02	3.70E+01	8.24E+01	6.73E-01
1.54E+02	3.76E+01	8.32E+01	6.79E-01
1.57E+02	3.82E+01	8.40E+01	6.86E-01

TABLE B-3 (Continued)

3.90E+01	5.05E-03	8.54E-01	0.00E+00	2.81E+01	7.54E+01	1.60E+02	3.90E+01	8.49E+01	6.93E-01
3.98E+01	4.99E-03	8.51E-01	0.00E+00	2.85E+01	7.63E+01	1.63E+02	3.98E+01	8.60E+01	7.02E-01
4.08E+01	4.92E-03	8.48E-01	0.00E+00	2.89E+01	7.72E+01	1.67E+02	4.08E+01	8.72E+01	7.12E-01
4.19E+01	4.84E-03	8.44E-01	0.00E+00	2.93E+01	7.83E+01	1.72E+02	4.19E+01	8.86E+01	7.23E-01
4.32E+01	4.75E-03	8.40E-01	0.00E+00	2.98E+01	7.94E+01	1.77E+02	4.32E+01	9.01E+01	7.36E-01
4.46E+01	4.65E-03	8.37E-01	0.00E+00	3.03E+01	8.08E+01	1.83E+02	4.46E+01	9.19E+01	7.51E-01
4.63E+01	4.53E-03	8.35E-01	0.00E+00	3.10E+01	8.23E+01	1.89E+02	4.63E+01	9.40E+01	7.67E-01
4.81E+01	4.40E-03	8.34E-01	0.00E+00	3.16E+01	8.39E+01	1.96E+02	4.81E+01	9.63E+01	7.87E-01
5.03E+01	4.26E-03	8.35E-01	0.00E+00	3.24E+01	8.57E+01	2.04E+02	5.03E+01	9.89E+01	8.08E-01
5.27E+01	4.10E-03	8.38E-01	0.00E+00	3.32E+01	8.78E+01	2.14E+02	5.27E+01	1.02E+02	8.33E-01
5.55E+01	3.93E-03	8.42E-01	0.00E+00	3.41E+01	9.00E+01	2.24E+02	5.55E+01	1.05E+02	8.61E-01
5.86E+01	3.74E-03	8.50E-01	0.00E+00	3.51E+01	9.24E+01	2.36E+02	5.86E+01	1.09E+02	8.93E-01
6.22E+01	3.54E-03	8.59E-01	0.00E+00	3.62E+01	9.51E+01	2.49E+02	6.22E+01	1.14E+02	9.29E-01
6.63E+01	3.34E-03	8.72E-01	0.00E+00	3.74E+01	9.79E+01	2.64E+02	6.63E+01	1.19E+02	9.71E-01
7.09E+01	3.12E-03	8.88E-01	0.00E+00	3.87E+01	1.01E+02	2.80E+02	7.09E+01	1.25E+02	1.02E+00
7.63E+01	2.91E-03	9.09E-01	0.00E+00	4.01E+01	1.04E+02	2.98E+02	7.63E+01	1.31E+02	1.07E+00
8.23E+01	2.69E-03	9.32E-01	0.00E+00	4.16E+01	1.08E+02	3.19E+02	8.23E+01	1.39E+02	1.13E+00
8.92E+01	2.48E-03	9.60E-01	0.00E+00	4.32E+01	1.12E+02	3.42E+02	8.92E+01	1.47E+02	1.20E+00
9.70E+01	2.27E-03	9.92E-01	0.00E+00	4.49E+01	1.16E+02	3.68E+02	9.70E+01	1.57E+02	1.28E+00
1.06E+02	2.06E-03	1.03E+00	0.00E+00	4.68E+01	1.20E+02	3.97E+02	1.06E+02	1.68E+02	1.37E+00
1.16E+02	1.87E-03	1.07E+00	0.00E+00	4.87E+01	1.24E+02	4.29E+02	1.16E+02	1.81E+02	1.48E+00
1.28E+02	1.68E-03	1.12E+00	0.00E+00	5.08E+01	1.29E+02	4.64E+02	1.28E+02	1.95E+02	1.59E+00
1.41E+02	1.51E-03	1.17E+00	0.00E+00	5.30E+01	1.34E+02	5.04E+02	1.41E+02	2.12E+02	1.73E+00
1.56E+02	1.34E-03	1.23E+00	0.00E+00	5.53E+01	1.39E+02	5.47E+02	1.56E+02	2.30E+02	1.88E+00
1.73E+02	1.19E-03	1.30E+00	0.00E+00	5.77E+01	1.44E+02	5.96E+02	1.73E+02	2.52E+02	2.06E+00
1.93E+02	1.06E-03	1.38E+00	0.00E+00	6.03E+01	1.50E+02	6.50E+02	1.93E+02	2.76E+02	2.25E+00
2.15E+02	9.31E-04	1.46E+00	0.00E+00	6.31E+01	1.55E+02	7.09E+02	2.15E+02	3.04E+02	2.48E+00
2.40E+02	8.18E-04	1.55E+00	0.00E+00	6.60E+01	1.61E+02	7.75E+02	2.40E+02	3.35E+02	2.74E+00
2.69E+02	7.17E-04	1.65E+00	0.00E+00	6.91E+01	1.68E+02	8.47E+02	2.69E+02	3.71E+02	3.03E+00
3.02E+02	6.26E-04	1.76E+00	0.00E+00	7.24E+01	1.74E+02	9.27E+02	3.02E+02	4.12E+02	3.36E+00
3.40E+02	5.46E-04	1.89E+00	0.00E+00	7.60E+01	1.81E+02	1.02E+03	3.40E+02	4.59E+02	3.74E+00
3.82E+02	4.76E-04	2.02E+00	0.00E+00	7.98E+01	1.88E+02	1.11E+03	3.82E+02	5.12E+02	4.18E+00
4.31E+02	4.13E-04	2.17E+00	0.00E+00	8.38E+01	1.95E+02	1.22E+03	4.31E+02	5.72E+02	4.67E+00
4.86E+02	3.59E-04	2.32E+00	0.00E+00	8.82E+01	2.03E+02	1.34E+03	4.86E+02	6.41E+02	5.23E+00
5.49E+02	3.11E-04	2.49E+00	0.00E+00	9.29E+01	2.11E+02	1.47E+03	5.49E+02	7.19E+02	5.87E+00
6.21E+02	2.70E-04	2.68E+00	0.00E+00	9.79E+01	2.20E+02	1.61E+03	6.21E+02	8.09E+02	6.60E+00
7.03E+02	2.33E-04	2.88E+00	0.00E+00	1.03E+02	2.29E+02	1.77E+03	7.03E+02	9.11E+02	7.43E+00
7.96E+02	2.02E-04	3.10E+00	0.00E+00	1.09E+02	2.39E+02	1.94E+03	7.96E+02	1.03E+03	8.38E+00
9.03E+02	1.75E-04	3.33E+00	0.00E+00	1.16E+02	2.49E+02	2.14E+03	9.03E+02	1.16E+03	9.46E+00
1.02E+03	1.51E-04	3.58E+00	0.00E+00	1.23E+02	2.59E+02	2.35E+03	1.02E+03	1.31E+03	1.07E+01

1.16E+03	1.31E-04	2.70E+02	1.30E+02	0.00E+00	3.86E+00	2.58E+03	1.16E+03	1.48E+03	1.21E+01
1.32E+03	1.13E-04	2.82E+02	1.39E+02	0.00E+00	4.15E+00	2.84E+03	1.32E+03	1.68E+03	1.37E+01
1.50E+03	9.81E-05	2.95E+02	1.48E+02	0.00E+00	4.46E+00	3.13E+03	1.50E+03	1.90E+03	1.55E+01
1.70E+03	8.49E-05	3.08E+02	1.58E+02	0.00E+00	4.80E+00	3.44E+03	1.70E+03	2.15E+03	1.76E+01
1.93E+03	7.35E-05	3.22E+02	1.69E+02	0.00E+00	5.16E+00	3.79E+03	1.93E+03	2.44E+03	1.99E+01
2.20E+03	6.37E-05	3.37E+02	1.80E+02	0.00E+00	5.71E+00	3.80E+03	2.20E+03	2.44E+03	4.18E+02
2.53E+03	5.53E-05	3.53E+02	1.94E+02	0.00E+00	6.31E+00	4.02E+03	2.53E+03	2.45E+03	6.25E+02

time averaged (tav = 600. s) volume concentration: concentration in the z = 0.00 plane.

average volume concentration at (x,y,z)

downwind distance x (m)	time of peak conc. (s)	cloud duration (s)	effective half width bbc (m)	yfl	y/bbc= 0.0	y/bbc= 0.5	y/bbc= 1.0	y/bbc= 1.5	y/bbc= 2.0	y/bbc= 2.5
-3.65E+01	1.83E+03	3.60E+03	3.65E+01	0.	0.00E+00	0.00E+00	0.00E+00	0.00E+00	0.00E+00	0.00E+00
-2.92E+01	1.83E+03	3.60E+03	3.71E+01	59.	8.29E-03	7.75E-03	2.82E-03	8.16E-05	8.77E-08	0.00E+00
-2.19E+01	1.82E+03	3.60E+03	3.79E+01	63.	1.49E-02	1.38E-02	5.01E-03	1.61E-04	2.29E-07	0.00E+00
-1.46E+01	1.81E+03	3.60E+03	3.95E+01	67.	2.09E-02	1.92E-02	6.91E-03	2.48E-04	4.63E-07	0.00E+00
-7.29E+00	1.81E+03	3.60E+03	4.23E+01	73.	2.63E-02	2.39E-02	8.58E-03	3.42E-04	8.27E-07	0.00E+00
-8.58E-06	1.80E+03	3.60E+03	4.67E+01	82.	3.11E-02	2.80E-02	1.00E-02	4.38E-04	1.34E-06	0.00E+00
7.29E+00	1.81E+03	3.60E+03	5.26E+01	94.	3.51E-02	3.13E-02	1.12E-02	5.29E-04	1.97E-06	0.00E+00
1.46E+01	1.81E+03	3.60E+03	6.00E+01	108.	3.82E-02	3.38E-02	1.20E-02	6.08E-04	2.67E-06	1.14E-09
2.19E+01	1.82E+03	3.60E+03	6.88E+01	125.	4.04E-02	3.54E-02	1.26E-02	6.73E-04	3.41E-06	1.21E-09
2.92E+01	1.83E+03	3.60E+03	7.78E+01	142.	4.13E-02	3.60E-02	1.28E-02	7.10E-04	3.94E-06	2.48E-09
3.65E+01	1.83E+03	3.60E+03	8.68E+01	159.	4.15E-02	3.60E-02	1.27E-02	7.29E-04	4.35E-06	2.49E-09
3.70E+01	1.83E+03	3.60E+03	8.74E+01	161.	4.12E-02	3.57E-02	1.26E-02	7.25E-04	4.34E-06	2.47E-09
3.76E+01	1.83E+03	3.60E+03	8.81E+01	162.	4.08E-02	3.54E-02	1.25E-02	7.20E-04	4.33E-06	2.45E-09
3.82E+01	1.84E+03	3.60E+03	8.89E+01	163.	4.05E-02	3.51E-02	1.24E-02	7.15E-04	4.32E-06	2.43E-09
3.90E+01	1.84E+03	3.60E+03	8.98E+01	165.	4.00E-02	3.47E-02	1.23E-02	7.09E-04	4.31E-06	2.40E-09
3.98E+01	1.84E+03	3.60E+03	9.08E+01	167.	3.95E-02	3.43E-02	1.21E-02	7.02E-04	4.30E-06	2.37E-09
4.08E+01	1.84E+03	3.60E+03	9.20E+01	169.	3.90E-02	3.38E-02	1.20E-02	6.94E-04	4.27E-06	2.34E-09
4.19E+01	1.84E+03	3.60E+03	9.33E+01	171.	3.84E-02	3.32E-02	1.18E-02	6.84E-04	4.25E-06	2.30E-09
4.32E+01	1.84E+03	3.60E+03	9.47E+01	174.	3.76E-02	3.26E-02	1.15E-02	6.73E-04	4.21E-06	2.26E-09
4.46E+01	1.84E+03	3.60E+03	9.64E+01	176.	3.68E-02	3.19E-02	1.13E-02	6.60E-04	4.17E-06	2.21E-09
4.63E+01	1.84E+03	3.60E+03	9.82E+01	180.	3.59E-02	3.11E-02	1.10E-02	6.46E-04	4.12E-06	2.16E-09
4.81E+01	1.84E+03	3.60E+03	1.00E+02	183.	3.49E-02	3.01E-02	1.07E-02	6.29E-04	4.05E-06	2.10E-09
5.03E+01	1.85E+03	3.60E+03	1.02E+02	187.	3.37E-02	2.91E-02	1.03E-02	6.10E-04	3.97E-06	2.03E-09
5.27E+01	1.85E+03	3.60E+03	1.05E+02	191.	3.25E-02	2.80E-02	9.90E-03	5.90E-04	3.88E-06	1.95E-09
5.55E+01	1.85E+03	3.60E+03	1.08E+02	196.	3.11E-02	2.68E-02	9.47E-03	5.67E-04	3.77E-06	2.80E-09
5.86E+01	1.85E+03	3.60E+03	1.11E+02	201.	2.96E-02	2.55E-02	9.01E-03	5.42E-04	3.65E-06	2.67E-09
6.22E+01	1.86E+03	3.60E+03	1.14E+02	206.	2.80E-02	2.41E-02	8.53E-03	5.16E-04	3.52E-06	2.53E-09
6.63E+01	1.86E+03	3.60E+03	1.17E+02	212.	2.64E-02	2.27E-02	8.02E-03	4.88E-04	3.37E-06	2.38E-09

TABLE B-3 (Continued)

7.09E+01	1.87E+03	3.60E+03	1.21E+02	219.	2.47E-02	2.12E-02	7.50E-03	4.59E-04	3.22E-06	2.23E-09
7.63E+01	1.87E+03	3.60E+03	1.25E+02	225.	2.30E-02	1.97E-02	6.97E-03	4.29E-04	3.06E-06	2.08E-09
8.23E+01	1.88E+03	3.60E+03	1.30E+02	232.	2.13E-02	1.82E-02	6.44E-03	3.99E-04	2.90E-06	1.92E-09
8.92E+01	1.88E+03	3.60E+03	1.34E+02	240.	1.96E-02	1.68E-02	5.91E-03	3.69E-04	2.73E-06	1.77E-09
9.70E+01	1.89E+03	3.60E+03	1.39E+02	248.	1.79E-02	1.53E-02	5.40E-03	3.40E-04	2.56E-06	2.16E-09
1.06E+02	1.90E+03	3.60E+03	1.45E+02	256.	1.63E-02	1.39E-02	4.90E-03	3.11E-04	2.40E-06	1.96E-09
1.16E+02	1.91E+03	3.60E+03	1.50E+02	264.	1.47E-02	1.26E-02	4.43E-03	2.84E-04	2.23E-06	1.78E-09
1.28E+02	1.92E+03	3.60E+03	1.56E+02	271.	1.33E-02	1.13E-02	3.98E-03	2.57E-04	2.07E-06	2.00E-09
1.41E+02	1.93E+03	3.60E+03	1.62E+02	282.	1.19E-02	1.01E-02	3.55E-03	2.32E-04	1.92E-06	1.79E-09
1.56E+02	1.95E+03	3.60E+03	1.69E+02	292.	1.06E-02	8.98E-03	3.16E-03	2.09E-04	1.77E-06	1.60E-09
1.73E+02	1.96E+03	3.60E+03	1.75E+02	302.	9.40E-03	7.96E-03	2.80E-03	1.87E-04	1.63E-06	1.70E-09
1.93E+02	1.98E+03	3.60E+03	1.83E+02	312.	8.31E-03	7.02E-03	2.47E-03	1.67E-04	1.50E-06	1.51E-09
2.15E+02	2.00E+03	3.60E+03	1.90E+02	322.	7.32E-03	6.16E-03	2.17E-03	1.49E-04	1.38E-06	1.55E-09
2.40E+02	2.02E+03	3.60E+03	1.98E+02	333.	6.42E-03	5.40E-03	1.89E-03	1.32E-04	1.27E-06	1.36E-09
2.69E+02	2.05E+03	3.60E+03	2.06E+02	344.	5.62E-03	4.71E-03	1.65E-03	1.17E-04	1.16E-06	1.34E-09
3.02E+02	2.08E+03	3.60E+03	2.15E+02	356.	4.90E-03	4.10E-03	1.44E-03	1.03E-04	1.07E-06	1.42E-09
3.40E+02	2.12E+03	3.60E+03	2.24E+02	368.	4.27E-03	3.55E-03	1.24E-03	9.08E-05	9.82E-07	1.35E-09
3.82E+02	2.16E+03	3.60E+03	2.33E+02	380.	3.71E-03	3.08E-03	1.08E-03	7.99E-05	9.02E-07	1.37E-09
4.31E+02	2.20E+03	3.60E+03	2.43E+02	393.	3.21E-03	2.66E-03	9.29E-04	7.03E-05	8.30E-07	1.36E-09
4.86E+02	2.25E+03	3.60E+03	2.54E+02	407.	2.78E-03	2.29E-03	8.01E-04	6.17E-05	7.65E-07	1.32E-09
5.49E+02	2.31E+03	3.60E+03	2.66E+02	420.	2.41E-03	1.97E-03	6.89E-04	5.42E-05	7.05E-07	1.33E-09
6.21E+02	2.38E+03	3.60E+03	2.78E+02	435.	2.08E-03	1.70E-03	5.91E-04	4.75E-05	6.51E-07	1.37E-09
7.03E+02	2.45E+03	3.60E+03	2.91E+02	450.	1.79E-03	1.46E-03	5.07E-04	4.17E-05	6.02E-07	1.38E-09
7.96E+02	2.54E+03	3.60E+03	3.05E+02	466.	1.55E-03	1.25E-03	4.35E-04	3.65E-05	5.58E-07	1.39E-09
9.03E+02	2.64E+03	3.60E+03	3.19E+02	482.	1.33E-03	1.07E-03	3.72E-04	3.20E-05	5.17E-07	1.45E-09
1.02E+03	2.75E+03	3.60E+03	3.35E+02	499.	1.15E-03	9.16E-04	3.18E-04	2.80E-05	4.80E-07	1.47E-09
1.16E+03	2.88E+03	3.60E+03	3.52E+02	517.	9.85E-04	7.83E-04	2.71E-04	2.45E-05	4.45E-07	1.50E-09
1.32E+03	3.03E+03	3.60E+03	3.71E+02	536.	8.47E-04	6.69E-04	2.32E-04	2.14E-05	4.13E-07	1.55E-09
1.50E+03	3.19E+03	3.60E+03	3.91E+02	555.	7.27E-04	5.71E-04	1.97E-04	1.87E-05	3.84E-07	1.57E-09
1.70E+03	3.38E+03	3.60E+03	4.12E+02	575.	6.24E-04	4.87E-04	1.68E-04	1.63E-05	3.56E-07	1.60E-09
1.93E+03	3.60E+03	3.60E+03	4.35E+02	595.	5.35E-04	4.15E-04	1.43E-04	1.42E-05	3.29E-07	1.65E-09
2.20E+03	3.80E+03	3.61E+03	4.60E+02	616.	4.59E-04	3.54E-04	1.22E-04	1.24E-05	3.05E-07	1.70E-09
2.53E+03	4.02E+03	3.64E+03	4.87E+02	638.	3.94E-04	3.01E-04	1.03E-04	1.08E-05	2.84E-07	

1

time averaged (tav = 600. s) volume concentration: maximum concentration (volume fraction) along centerline.

downwind distance x (m)	height z (m)	maximum concentration c(x,0,z)	time of max conc (s)	cloud duration (s)
-3.65E+01	0.00E+00	0.00E+00	1.83E+03	3.60E+03
-2.92E+01	0.00E+00	8.29E-03	1.83E+03	3.60E+03
-2.19E+01	0.00E+00	1.49E-02	1.82E+03	3.60E+03
-1.46E+01	0.00E+00	2.09E-02	1.81E+03	3.60E+03
-7.29E+00	0.00E+00	2.63E-02	1.81E+03	3.60E+03
-8.58E-06	0.00E+00	3.11E-02	1.80E+03	3.60E+03
7.29E+00	0.00E+00	3.51E-02	1.81E+03	3.60E+03
1.46E+01	0.00E+00	3.82E-02	1.81E+03	3.60E+03
2.19E+01	0.00E+00	4.04E-02	1.82E+03	3.60E+03
2.92E+01	0.00E+00	4.13E-02	1.83E+03	3.60E+03
3.65E+01	0.00E+00	4.15E-02	1.83E+03	3.60E+03
3.70E+01	0.00E+00	4.12E-02	1.83E+03	3.60E+03
3.76E+01	0.00E+00	4.08E-02	1.83E+03	3.60E+03
3.82E+01	0.00E+00	4.05E-02	1.84E+03	3.60E+03
3.90E+01	0.00E+00	4.00E-02	1.84E+03	3.60E+03
3.98E+01	0.00E+00	3.95E-02	1.84E+03	3.60E+03
4.08E+01	0.00E+00	3.90E-02	1.84E+03	3.60E+03
4.19E+01	0.00E+00	3.84E-02	1.84E+03	3.60E+03
4.32E+01	0.00E+00	3.76E-02	1.84E+03	3.60E+03
4.46E+01	0.00E+00	3.68E-02	1.84E+03	3.60E+03
4.63E+01	0.00E+00	3.59E-02	1.84E+03	3.60E+03
4.81E+01	0.00E+00	3.49E-02	1.84E+03	3.60E+03
5.03E+01	0.00E+00	3.37E-02	1.85E+03	3.60E+03
5.27E+01	0.00E+00	3.25E-02	1.85E+03	3.60E+03
5.55E+01	0.00E+00	3.11E-02	1.85E+03	3.60E+03
5.86E+01	0.00E+00	2.96E-02	1.86E+03	3.60E+03
6.22E+01	0.00E+00	2.80E-02	1.86E+03	3.60E+03
6.63E+01	0.00E+00	2.64E-02	1.87E+03	3.60E+03
7.09E+01	0.00E+00	2.47E-02	1.87E+03	3.60E+03
7.63E+01	0.00E+00	2.30E-02	1.88E+03	3.60E+03
8.23E+01	0.00E+00	2.13E-02	1.88E+03	3.60E+03
8.92E+01	0.00E+00	1.96E-02	1.89E+03	3.60E+03
9.70E+01	0.00E+00	1.79E-02	1.89E+03	3.60E+03
1.06E+02	0.00E+00	1.63E-02	1.90E+03	3.60E+03
1.16E+02	0.00E+00	1.47E-02	1.91E+03	3.60E+03
1.28E+02	0.00E+00	1.33E-02	1.92E+03	3.60E+03
1.41E+02	0.00E+00	1.19E-02	1.93E+03	3.60E+03
1.56E+02	0.00E+00	1.06E-02	1.95E+03	3.60E+03
1.73E+02	0.00E+00	9.40E-03	1.96E+03	3.60E+03
1.93E+02	0.00E+00	8.31E-03	1.98E+03	3.60E+03
2.15E+02	0.00E+00	7.32E-03	2.00E+03	3.60E+03
2.40E+02	0.00E+00	6.42E-03	2.02E+03	3.60E+03
2.69E+02	0.00E+00	5.62E-03	2.05E+03	3.60E+03
3.02E+02	0.00E+00	4.90E-03	2.08E+03	3.60E+03
3.40E+02	0.00E+00	4.27E-03	2.12E+03	3.60E+03
3.82E+02	0.00E+00	3.71E-03	2.16E+03	3.60E+03
4.31E+02	0.00E+00	3.21E-03	2.20E+03	3.60E+03
4.86E+02	0.00E+00	2.78E-03	2.25E+03	3.60E+03
5.49E+02	0.00E+00	2.41E-03	2.31E+03	3.60E+03
6.21E+02	0.00E+00	2.08E-03	2.38E+03	3.60E+03
7.03E+02	0.00E+00	1.79E-03	2.45E+03	3.60E+03
7.96E+02	0.00E+00	1.55E-03	2.54E+03	3.60E+03
9.03E+02	0.00E+00	1.33E-03	2.64E+03	3.60E+03
1.02E+03	0.00E+00	1.15E-03	2.75E+03	3.60E+03
1.16E+03	0.00E+00	9.85E-04	2.88E+03	3.60E+03
1.32E+03	0.00E+00	8.47E-04	3.03E+03	3.60E+03
1.50E+03	0.00E+00	7.27E-04	3.19E+03	3.60E+03
1.70E+03	0.00E+00	6.24E-04	3.38E+03	3.60E+03
1.93E+03	0.00E+00	5.35E-04	3.60E+03	3.60E+03
2.20E+03	0.00E+00	4.59E-04	3.80E+03	3.61E+03
2.53E+03	0.00E+00	3.94E-04	4.02E+03	3.64E+03

APPENDIX C

Scenario 3: Liquid (Noncryogenic) Spill of Liquid Acetone

Scenario 3 is intended to represent noncryogenic (i.e., nonboiling) liquid spills. This scenario contrasts with Scenario 2, where the liquid spill of chlorine was cryogenic. A spill of liquid acetone (C_3H_6O) is assumed for Scenario 3, where the liquid pours out of a 0.0508-m (2-in) nozzle connection near the base of a vertical cylindrical storage tank of height 7.62 m (25 ft) and diameter 7.62 m (25 ft). Figure C-1 contains a schematic diagram illustrating Scenario 3. The liquid acetone spills into a concrete diked area of height 0.914 m (3 ft) and diameter 21.9 m (72 ft). The ambient atmosphere is assumed to be neutral (class D) with a wind speed of 5 m s–1 measured at a height of 10 m. The acetone is assumed to be stored at ambient temperature (303 K) and pressure (1 atm). Tables 9-2 and C-1 list the properties of acetone, the source configuration, the meteorological data, and the site conditions for Scenario 3.

Since the boiling point temperature (329.4 K) of acetone is 26.4 K higher than the ambient temperature, the spill is noncryogenic. Because this is a simpler spill scenario than Scenario 2, it is possible to use analytical

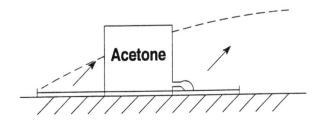

Figure C-1. Schematic diagram of Scenario 3, a liquid spill of acetone into a diked area.

TABLE C-1
Input Data Archive for Scenario 3, a Liquid Spill of Acetone into a Diked Area
(Data Taken from Table 9-2, which covers All Seven Scenarios)

ACETONE PROPERTIES	
molecular weight	$M = 58.1$ kg kmole^{-1}
normal boiling point	$T_b = 329.4$ K
latent heat of evaporation	$H_{vap} = 545,000$ J kg^{-1}
specific heat at constant pressure for vapor	$c_{pv} = 1467$ J kg^{-1} K^{-1}
specific heat at constant pressure for liquid	$c_{pl} = 2070$ J kg^{-1} K^{-1}
density of liquid	$\rho_l = 788$ kg m^{-3}
molecular diffusivity	$D_m = 1.23 \times 10^{-5}$ m^2 s^{-1}
kinematic viscosity	$\nu = 1.10 \times 10^{-5}$ m^2 s^{-1}
SOURCE CONFIGURATION	
storage temperature	$T_s = 303$ K
release temperature	$T_o = 303$ K (normal boiling point)
storage pressure	$p_s = 1$ atm
pipe diameter	$D_p = 0.0508$ m
source height	$h_s = 1.0$ m
mass in tank	Mass = 143,743 kg
tank diameter	$D_t = 7.62$ m
tank height	$h_t = 7.62$ m
fluid level	$h_f = 4$ m
ground surface	concrete
source strength	$Q = 2.24$ kg s^{-1}
dike diameter	$D_o = 21.9$ m
METEOROLOGICAL CONDITIONS	
temperature	$T_a = 303$ K at height of 10 m
wind speed	$u = 5$ m s^{-1} at height of 10 m
stability class	D
pressure	$p_a = 1$ atm
relative humidity	RH = 50%
Monin-Obukhov length	$L = 9999$ m
friction velocity	$u_* = 0.344$ m s^{-1}
surface roughness	$z_0 = 0.03$ m (uncut grass)
SENSITIVITY RUNS	
RUN 1	Change stability class to F (stable) and decrease wind speed to 2 ms^{-1}
RUN 2	Double dike diameter from 21.9 to 43.8 m
RUN 3	Change averaging time T_a from 600 s to 0 s

CONCENTRATIONS OF INTEREST			
averaging time	$T_a = 600$ s	IDLH	$C = 20,000$ ppm
UEL	$C = 128,000$ ppm	LEL	$C = 26,000$ ppm

formulas to illustrate how the solutions can be obtained. The Fleischer equation (4-29) used in the Shell SPILLS model can be applied to estimate the evaporation rate, assuming steady-state conditions:

$$Q = \frac{k_g A_p p_s M}{RT_a}$$ (C-1)

where

Q = evaporation or gas mass emission rate (kg s^{-1})

k_g = mass transfer coefficient (m s^{-1})

A_p = pool area (m^2)

p_s = vapor pressure of the compound at temperature T (Pa)

M = molecular weight (kg kmole^{-1})

R = gas constant (8.31 J mole^{-1} K^{-1})

T_a = ambient temperature (K)

The pool area, A_p, for Scenario 3, is the area enclosed by the dike (376.68 m^2) minus the area covered by the tank (45.60 m^2), or 331.08 m^2. This area corresponds to a circle with an effective radius of 10.27 m. The saturation vapor pressure, p_s, for acetone at 303 K is 37932 Pa (about 0.38 atm). The molecular weight, M, for acetone is 58.1 kg kmole^{-1}. The chemical engineering approach given by equation (4-30) is used to calculate the mass transfer coefficient, k_g, where it is necessary to specify the Reynolds number, N_{Re}, the Schmidt number, N_{Sc}, and the Sherwood number, N_{Sh}. N_{Re} can be calculated from the formula:

$$N_{Re} = \frac{u d_p}{\nu}$$ (C-2)

where

u = ambient wind speed = 5 m s^{-1}

d_p = pool diameter = 20.54 m

ν = kinematic viscosity = 1.10 × 10^{-5} m^2 s^{-1} for acetone

Thus, N_{Re} = 9.336 × 10^6. The Schmidt number, N_{Sc}, can be calculated from the formula:

$$N_{Sc} = \frac{\nu}{D_m}$$ (C-3)

where D_m is the molecular diffusivity = 1.23 × 10^{-5} m^2 s^{-1} for acetone.

Thus, $N_{Sc} = 0.8943$. The Sherwood number, N_{Sh}, can be calculated from equation (4-32):

$$N_{Sh} = 0.037\, N_{Sc}^{\frac{1}{3}}[N_{Re}^{0.8} - 15{,}200] = 12{,}900 \qquad (C\text{-}4)$$

The mass transfer coefficient, k_g, can then be calculated from equation (4-30):

$$k_g = \frac{N_{Sh}D_m}{d_p} = 7.725 \times 10^{-3} \ \text{m s}^{-1} \qquad (C\text{-}5)$$

Therefore, the evaporation rate or the gas mass emission rate, Q, given by equation (C-1) equals 2.24 kg s^{-1}, which is listed in Tables 9-2 and C-1.

The DEGADIS and HGSYSTEM models are applied to this scenario using the evaporation rate, 2.24 kg s^{-1}, calculated above. However, the ALOHA model is applied in such a way that it calculates its own evaporative source term, which yields Q = 1.27 kg s^{-1}. The ALOHA-calculated evaporation rate is 43% less than the 2.24 kg s^{-1} value calculated by the analytical equations (C-2) through (C-5). We recommend using the conservative (larger) value of 2.24 kg s^{-1} in worst-case analyses.

Before applying any dense gas dispersion models to this scenario, it is advisable to first calculate the source Richardson number and determine if it exceeds the critical value of 50. Equation (5-1) is appropriate for continuous ground-level releases from area sources:

$$Ri_o = \frac{g(\rho_{po} - \rho_a)V_{co}}{\rho_a D_o u_*^3} \qquad (C\text{-}6)$$

The following are prescribed as input in Table C-1 or can be calculated from known parameters:

ρ_{po} = 2.34 kg m^{-3}

ρ_a = 1.16 kg m^{-3}

g = 9.8 kg m^{-3}

u_* = 0.344 m s^{-1}

V_{co} = $Q/(\pi\rho_{po}) = 0.305$ m^3 s^{-1}

D_o = 2 · 10.27 = 20.54 m

By substituting these variables into equation (C-6), it is found that Ri_o = 3.61, which is a factor of about 15 less than the critical Ri_o. It is concluded that the dense gas effects are probably not very great in this scenario, and it may be more appropriate to apply a passive gas model such as ISC (EPA 1995). However, because one of the purposes of this *Guidelines* book is to demonstrate the use of models such as ALOHA, DEGADIS, and

HGSYSTEM, and there are likely to be some dense gas effects near the source, we proceed with the application of those models.

The ALOHA, DEGADIS and HGSYSTEM vapor cloud dispersion models directly treat ground-level area-source releases such as the evaporation from the liquid pool in Scenario 3. Both DEGADIS and HGSYSTEM are applied in steady-state mode with a source strength of 2.24 kg s^{-1} and, as mentioned above, ALOHA is applied with an evaporative source strength of 1.27 kg s^{-1} (calculated internally). It should be noted that the HEGADAS module in HGSYSTEM has been applied to this case.

The three models were each applied to a base case and to three sensitivity cases. The changes to the input conditions, involving the stability class/wind speed, the pool diameter, and the averaging time for the sensitivity runs are listed near the end of Table C-1. The first run is called the "base run" for each model. Predicted ground-level concentrations, C, as functions of downwind distance, x, are listed in Table C-2 and are plotted in Figures C-2 and C-3. Figure C-2 shows that the concentration curves predicted by the three models have similar slopes and are all within a factor of four of each other. DEGADIS tends to predict the highest concentrations and ALOHA predicts the lowest concentrations at each downwind distance, with the HGSYSTEM predictions only 10% or 20% less than the DEGADIS predictions. A major part of the reason for the relatively low ALOHA predictions is likely that its evaporative source term is 43% less than the source term for DEGADIS and HGSYSTEM. However, even if the ALOHA source term were adjusted upward by 43%, the model would still underpredict by a factor of about two relative to DEGADIS. Note that the slope of the DEGADIS $C(x)$ curve is much less for Scenario 3 than for Scenario 2, due to the model's tendency to simulate a "vapor blanket" over the source in Scenario 2 but not in Scenario 3. The presence of vapor blankets leads to high predicted concentrations near the source and rapid decreases in predicted concentrations at moderate distances. Vapor blankets are less likely to be generated by the models during neutral conditions with moderate winds and smaller mass emission rates.

The results of the sensitivity runs for DEGADIS are plotted in Figure C-3. It is seen that when stability class/wind speed changes from D/5 to F/2, predicted concentrations increase by a factor of two to four and there is a hint of a vapor blanket effect at near distances. When the dike (pool) diameter is doubled, the mass emission rate and hence the predicted concentrations increase by about a factor of four. When the averaging time, T_a, is decreased from 600 s to 0 s, the predicted concentrations are not changed much at x less than 250 m, but increase by a factor of two at large

TABLE C-2
Predicted Cloud Centerline Concentration (ppm) as a Function of Downwind
Distance for ALOHA, DEGADIS, and HGSYSTEM for Base Run and Three
Sensitivity Runs for Scenario 3 (Acetone Spill)

| | | SCENARIO 3 | | |
| | | Centerline Concentration (ppm) | | |
		ALOHA	DEGADIS	HGSYSTEM
Base Run:	100 m	655	2585	1916
D/5	200 m	232	935	599
D_d=21.9m	500 m	48	160	109
T_a=600s	1000 m	15	40	29
	1500 m	7.8	18	13
	2000 m	5	11	7.5
Sensitivity Run 1:	100 m	3210	6832	8637
F/2	200 m	905	1437	3405
	500 m	184	282	1092
	1000 m	59	94	463
	1500 m	31	52	276
	2000 m	20	34	188
Sensitivity Run 2:	100 m	673	6347	5381
D_d=43.8m	200 m	247	2380	1976
	500 m	52	611	416
	1000 m	16	180	114
	1500 m	8.4	78	52
	2000 m	5.4	43	30.1
Sensitivity Run 3:	100 m	655	2512	2340
T_a=0s	200 m	232	936	886
	500 m	48	236	194
	1000 m	15	84	55
	1500 m	7.8	42	26
	2000 m	5	23	14.8

distances. This discrepancy is caused by the assumption within DEGADIS
that averaging time does not influence the dense gas core of the cloud, which
affects the concentration predictions near the source.

As an example of model input/output files, the hard copy DEGADIS
output file for the base run of Scenario 3 is listed in the next few pages as
Table C-3. These files as well as similar ALOHA and HGSYSTEM files
for all runs are provided in the diskette in the pocket inside the back cover.

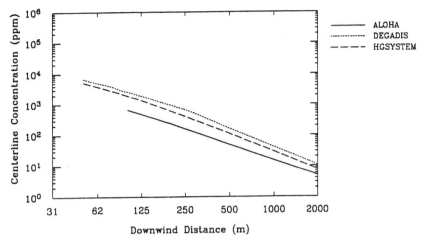

Figure C-2. Predicted cloud centerline concentrations as a function of downwind distance for Scenario 3 (acetone spill) base run for ALOHA, DEGADIS, and HGSYSTEM.

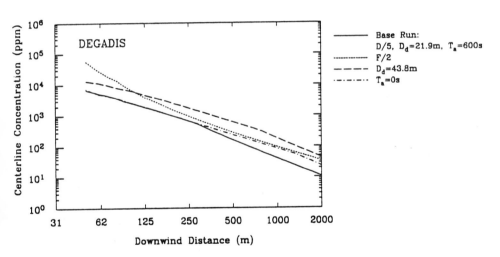

Figure C-3. Predicted cloud centerline concentrations as a function of downwind distance for Scenario 3 (acetone spill) for DEGADIS base run and three sensitivity runs.

TABLE C-3 Output File for DEGADIS for Scenario 3 Base Run

```
Data input on                        NO CREATION DATE
Source program run on                8-AUG-1995 17:14:16.68

o    TITLE BLOCK

TRIAL AND EXPERIMENT: S3A        Scenario 3
CHEMICAL: Acetone
STEADY-STATE SIMULATION IN ISOTHERMAL MODE


     Wind velocity at reference height                    5.00    m/s
     Reference height                                    10.00    m
o    Surface roughness length                         3.000E-02   m
o    Pasquill Stability class                                D
o    Monin-Obukhov length                             9.999E+03   m
     Gaussian distribution constants
                   Specified averaging time          600.00      s
                                        Delta          .13600
                                        Beta           .90000
                                        Alpha          .22698
o    Wind velocity power law constant                    .30085   m/s
o    Friction velocity                                303.00      K
     Ambient Temperature                                1.000      atm
     Ambient Pressure                                1.377E-02    kg/kg BDA
     Ambient Absolute Humidity                          50.00      %
     Ambient Relative Humidity


     Input:        Mole fraction      CONCENTRATION OF C        GAS DENSITY
                                           kg/m**3               kg/m**3
                      .00000                .00000               1.15678
                     1.00000               2.33866               2.33866


o    Specified Gas Properties:

     Molecular weight:                                         58.100
     Release temperature:                                     303.00      K
     Density at release temperature and ambient pressure:      2.3387     kg/m**3
     Average heat capacity:                                   1467.0      J/kg K
     Upper mole fraction contour:                           2.00000E-02
     Lower mole fraction contour:                           1.00000E-07
     Height for isopleths:                                     .00000     m
```

Source input data points

Initial (pure contaminant) mass in cloud: .00000   kg

Time	Contaminant Mass Rate	Source Radius	Contaminant Mass Fraction	Temperature	Enthalpy
s	kg/s	m	kg contam/kg mix	K	J/kg
.00000	2.2400	10.265	1.0000	303.00	.00000
60230.	2.2400	10.265	1.0000	303.00	.00000
60231.	.00000	.00000	1.0000	303.00	.00000
60232.	.00000	.00000	1.0000	303.00	.00000

o   Calculation procedure for ALPHA: 1
o   Entrainment prescription for PHI: 3
o   Layer thickness ratio used for average depth: 2.1500
o   Air entrainment coefficient used: .590
o   Gravity slumping velocity coefficient used: 1.150
o   Isothermal calculation
o   Heat transfer not included
o   Water transfer not included

*****   CALCULATED SOURCE PARAMETERS   *****

Time sec	Gas Radius m	Height m	Qstar kg/m**2/s	SZ(x=L/2.) m	Mole frac C	Density kg/m**3	Rich No.
602.300	10.2650	.000000	6.766752E-03	1.07689	2.243641E-02	1.18331	.000000
1806.90	10.2650	.000000	6.766752E-03	1.07689	2.243641E-02	1.18331	.000000

0Source strength [kg/s] :   2.2400   20.530
Equivalent Primary source length [m] :   Equivalent Primary source radius [m] :   10.265
                                              Equivalent Primary source half-width [m] :   8.0621

Secondary source concentration [kg/m**3] : 5.24981E-02   Secondary source SZ [m] :   1.0769

Contaminant flux rate: 6.76675E-03

TABLE C-3 (Continued)

Secondary source mass fractions... contaminant: 4.436542E-02   air: .94265
Enthalpy: .00000   Density: 1.1833

Secondary source length [m] : 20.530   Secondary source half-width [m] : 8.0621

Distance (m)	Mole Fraction	Concentration (kg/m**3)	Density (kg/m**3)	Temperature (K)	Half Width (m)	Sz (m)	Sy (m)	Width at z= .00 m to: 1.000E-05mole% (m)	2.00 mole% (m)
10.3	2.244E-02	5.250E-02	1.18	303.	8.06	1.08	.000	8.06	8.06
10.4	2.236E-02	5.232E-02	1.18	303.	7.68	1.08	.453	9.27	7.83
10.6	2.215E-02	5.183E-02	1.18	303.	7.38	1.08	.846	10.4	7.66
12.0	2.087E-02	4.883E-02	1.18	303.	6.74	1.10	1.89	13.3	7.13
13.1	1.990E-02	4.657E-02	1.18	303.	6.47	1.12	2.45	15.0	
15.2	1.830E-02	4.282E-02	1.18	303.	6.16	1.16	3.25	17.5	
53.0	5.745E-03	1.344E-02	1.16	303.	5.01	2.00	11.1	41.6	
90.9	2.844E-03	6.655E-03	1.16	303.	4.32	2.89	16.5	57.2	
131.	1.704E-03	3.986E-03	1.16	303.	3.42	3.82	21.5	70.6	
171.	1.158E-03	2.710E-03	1.16	303.	2.41	4.71	26.1	82.2	
211.	8.509E-04	1.991E-03	1.16	303.	1.33	5.59	30.3	92.5	
251.	6.583E-04	1.540E-03	1.16	303.	.208	6.44	34.3	102.	
258.	6.350E-04	1.486E-03	1.16	303.	.000	6.59	34.8	103.	
318.	4.046E-04	9.468E-04	1.16	303.	.000	8.39	40.6	117.	
378.	2.819E-04	6.596E-04	1.16	303.	.000	10.1	46.3	130.	
438.	2.083E-04	4.875E-04	1.16	303.	.000	11.8	51.9	143.	
498.	1.606E-04	3.758E-04	1.16	303.	.000	13.4	57.5	156.	
558.	1.278E-04	2.990E-04	1.16	303.	.000	15.0	62.9	168.	
618.	1.042E-04	2.439E-04	1.16	303.	.000	16.6	68.4	180.	
678.	8.672E-05	2.029E-04	1.16	303.	.000	18.1	73.8	192.	
738.	7.333E-05	1.716E-04	1.16	303.	.000	19.6	79.1	203.	
798.	6.287E-05	1.471E-04	1.16	303.	.000	21.1	84.4	214.	
858.	5.452E-05	1.276E-04	1.16	303.	.000	22.5	89.7	225.	
918.	4.776E-05	1.117E-04	1.16	303.	.000	24.0	94.9	236.	

978.	4.219E-05	9.873E-05	1.16	303.	25.4	.000	100.	246.
1.038E+03	3.756E-05	8.789E-05	1.16	303.	26.8	.000	105.	256.
1.098E+03	3.366E-05	7.877E-05	1.16	303.	28.2	.000	110.	266.
1.158E+03	3.035E-05	7.102E-05	1.16	303.	29.6	.000	116.	276.
1.218E+03	2.751E-05	6.438E-05	1.16	303.	30.9	.000	121.	286.
1.278E+03	2.506E-05	5.864E-05	1.16	303.	32.3	.000	126.	295.
1.338E+03	2.292E-05	5.364E-05	1.16	303.	33.6	.000	131.	305.
1.398E+03	2.106E-05	4.927E-05	1.16	303.	34.9	.000	136.	314.
1.458E+03	1.941E-05	4.542E-05	1.16	303.	36.2	.000	141.	323.
1.518E+03	1.795E-05	4.201E-05	1.16	303.	37.5	.000	146.	332.
1.578E+03	1.665E-05	3.897E-05	1.16	303.	38.8	.000	151.	341.
1.638E+03	1.550E-05	3.626E-05	1.16	303.	40.1	.000	156.	349.
1.698E+03	1.446E-05	3.382E-05	1.16	303.	41.4	.000	161.	358.
1.758E+03	1.352E-05	3.163E-05	1.16	303.	42.6	.000	165.	367.
1.818E+03	1.267E-05	2.965E-05	1.16	303.	43.9	.000	170.	375.
1.878E+03	1.190E-05	2.785E-05	1.16	303.	45.1	.000	175.	383.
1.938E+03	1.120E-05	2.621E-05	1.16	303.	46.4	.000	180.	391.
1.998E+03	1.056E-05	2.471E-05	1.16	303.	47.6	.000	185.	399.
2.058E+03	9.976E-06	2.334E-05	1.16	303.	48.8	.000	190.	407.
2.118E+03	9.439E-06	2.209E-05	1.16	303.	50.1	.000	195.	415.
2.178E+03	8.944E-06	2.093E-05	1.16	303.	51.3	.000	199.	423.
...								
2.282E+04	1.008E-07	2.359E-07	1.16	303.	362.	.000	1.613E+03	146.
2.294E+04	9.982E-08	2.336E-07	1.16	303.	363.	.000	1.621E+03	

0

For the UFL of 2.0000 mole percent, and the LFL of 1.00000E-05 mole percent:

The mass of contaminant between the UFL and LFL is: 4139.2 kg.
The mass of contaminant above the LFL is: 4139.4 kg.

APPENDIX D

Scenario 4: Vertical Jet Release of a Dense Gas (Normal Butane)

Scenario 4 involves a jet release of a dense gas (normal butane, nC_4H_{10}) from a vertically oriented vent stack with inner diameter 0.75 m and elevation 5 m (see Figure D-1). The gas emission rate is assumed to equal 15 kg s^{-1}. The ambient atmosphere is assumed to be stable (class F) with a wind speed of 2 m s^{-1} at a reference height of 10 m. The initial jet temperature is assumed to equal the ambient temperature (303 K). The vent stack is assumed not to be under the influence of nearby buildings. Table 9-2 and Table D-1 list the properties of normal butane, the source configuration, the meteorological data, and the site conditions for Scenario 4.

No source model is applied to this scenario, since all source conditions are prescribed.

It is necessary to first calculate the initial Richardson number, Ri_o, in order to determine whether dense gas dispersion models are needed. The two definitions given by equations (5-3a) and (5-4b) are valid:

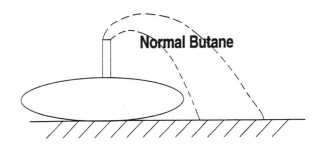

Figure D-1. Schematic diagram of Scenario 4, an elevated normal butane jet.

For maximum ground-level concentration at plume touchdown:

$$Ri_0 = \frac{g(\rho_{po} - \rho_a)V_{co}}{\rho_a u_*^3 h_s} \tag{D-1}$$

For jet trajectory very near source:

$$Ri_0 = \frac{g(\rho_{po} - \rho_a)V_{co}}{\rho_a u_*^3 D_p} \tag{D-2}$$

TABLE D-1
Input Data Archive for Scenario 4, a Vertical Gas Jet of Normal Butane
(Data Taken from Table 9-2, which Covers All Seven Scenarios)

NORMAL BUTANE PROPERTIES	
molecular weight	$M = 58.12$ kg kmole^{-1}
normal boiling point	$T_b = 272.7$ K
specific heat at constant pressure forvapor	$c_{pv} = 1715$ J kg^{-1} K^{-1}

SOURCE CONFIGURATION	
release temperature	$T_0 = 303$ K (normal boiling point)
stack diameter	$D_p = 0.75$ m
source height	$h_s = 5$ m
source strength	$Q = 15$ kg s^{-1} (steady state)

METEOROLOGICAL CONDITIONS	
temperature	$T_a = 303$ K at height of 10 m
wind speed	$u = 2$ m s^{-1} at height of 10 m
stability class	F
relative humidity	RH = 50%
Monin-Obukhov length	$L = 14.3$ m
friction velocity	$u_* = 0.099$ m s^{-1}
surface roughness	$z_0 = 0.1$ m scattered commercial/industrial

SENSITIVITY RUNS	
RUN 1	Increase stack height from 5 m to 20 m
RUN 2	Decrease stack diameter from 0.75 m to 0.30 m
RUN 3	Decrease averaging time from 600 s to 0 s

CONCENTRATIONS OF INTEREST			
averaging time	$T_a = 600$ s	IDLH	$C = 10,000$ ppm
UEL	$C = 84,100$ ppm	LEL	$C = 18,600$ ppm

The following variables are known from Table D-1 or are calculated from known parameters:

ρ_{po} = 2.34 kg m^{-3}

ρ_a = 1.16 kg m^{-3}

h_s = 5 m

u_* = 0.099 m s^{-1}

V_{co} = $Q/(\pi\rho_{po})$ = 2.04 m^3 s^{-1}

D_p = 0.75 m

Equations (D-1) and (D-2) then give Ri_o equal to 4220 and 28,100, respectively, both of which are several orders of magnitude above the critical Ri_o of 50. It is concluded that dense gas models are appropriate for Scenario 4, since the jet is dominated by density effects near the source and at downwind distances where jet touchdown occurs.

The DEGADIS and SLAB vapor cloud dispersion models have been applied to this Scenario. The Ooms dense gas jet module is applied by DEGADIS to the initial jet, since it has a vertical orientation. If the jet were not directed vertically (e.g., the horizontally directed jet in Scenario 1), DEGADIS would not use the Ooms jet model but would approximate the emission as a ground-level area source. SLAB directly treats jets with a vertical orientation, as well as jets at other angles.

Both DEGADIS and SLAB predict that the dense gas plume sinks to the ground within 25 m of the stack vent (see Figure D-2). This relatively rapid sinking of the plume is due to its large initial density. As seen in the figure, the dense jet is predicted to rise to a height of 13 to 15 m at a downwind distance of about 3 or 4 m before starting to sink to the ground. SLAB predicts that the plume centerline strikes the ground at a distance of 11 m, whereas DEGADIS predicts that the plume centerline strikes the ground farther out (25 m). Once the plume strikes the ground, both DEGADIS and SLAB treat its subsequent dispersion as if it were a ground-based area source.

Predicted cloud centerline concentrations and cloud half-widths at downwind distances from 50 m to 2000 m are listed in Table D-2 and predicted centerline concentrations are plotted in Figures D-3 and D-4. For the base run, the predicted DEGADIS and SLAB concentrations tend to agree within ±20% at distances from 50 m to 300 m, but the SLAB predictions tend to be higher than the DEGADIS predictions by a factor of four to six at distances from 1000 m to 2000 m (see Figure D-3). The cloud half-widths (lateral distances to the 10,000 ppm contour) are about 100 m (factor of 2) for both models at the 100 m and 200 m distances, but the

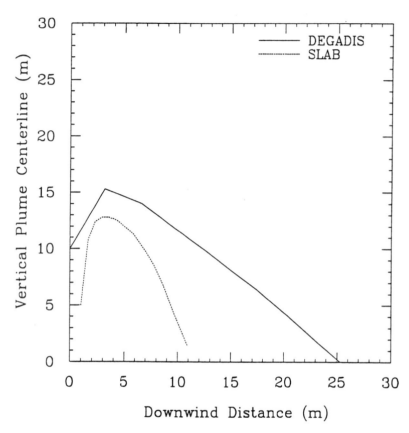

Figure D-2. Predicted plume centerline trajectories for Scenario 4 (normal butane jet) near the vent stack.

contour disappears for both models at greater distances. Figure D-3 shows that the downwind penetration of the 10,000 ppm contour is predicted to be about 250 m by both models.

The results of the sensitivity studies are plotted in Figure D-4 for the SLAB model. It is seen that the changes in stack height and averaging time have little effect on the predicted ground-level concentrations, since the plume quickly sinks to the ground and since the model's algorithms do not allow the averaging time to affect the dense gas core. The decrease in stack diameter from 0.75 m to 0.3 m causes a slight reduction in concentration of 20 to 30%, due to the increased initial plume speed and plume rise.

The sensitivity of the predicted distance where the bottom of the plume first sinks to the ground to changes in input parameters is given in Table D-3. The largest change occurs when the stack diameter is reduced from

TABLE D-2
Predicted Cloud Centerline Concentration (ppm) and Cloud Half-width (Lateral Distance to 10,000 ppm Contour) as a Function of Downwind Distance for DEGADIS and SLAB for the Base Run and Three Sensitivity Runs for Scenario 4 (Normal Butane Jet)

		SCENARIO 4			
		Centerline Concentration (ppm)		Lateral Distance to 10000 ppm (m)	
		DEGADIS	SLAB	DEGADIS	SLAB
Base Run:	100 m	27180	22620	59	99
h_s=5m	200 m	14935	11700	148	61
D_s=0.75m	500 m	1991	4384	n/a	n/a
T_a=600s	1000 m	451	1960	n/a	n/a
	1500 m	211	1207	n/a	n/a
	2000 m	128	857	n/a	n/a
Sensitivity Run 1:	100 m	12580	19040	n/a	87
h_s=20m	200 m	5217	10070	n/a	6
	500 m	2431	3956	n/a	n/a
	1000 m	650	1827	n/a	n/a
	1500 m	287	1147	n/a	n/a
	2000 m	166	820	n/a	n/a
Sensitivity Run 2:	100 m	11290	16700	n/a	44
D_s=0.3m	200 m	3604	6673	n/a	n/a
	500 m	1628	2701	n/a	n/a
	1000 m	757	1336	n/a	n/a
	1500 m	366	882	n/a	n/a
	2000 m	210	649	n/a	n/a
Sensitivity Run 3:	100 m	28165	22620	59	99
T_a=0s	200 m	15223	11800	153	61
	500 m	1971	4422	n/a	n/a
	1000 m	449	1997	n/a	n/a
	1500 m	210	1237	n/a	n/a
	2000 m	127	885	n/a	n/a

0.75 m to 0.30 m, which causes a factor of 6.25 increase in initial velocity and momentum flux and consequently a higher maximum plume rise. The change in stack height is seen to have a large effect on the DEGADIS predictions of distance (increase of a factor of four) but a much smaller

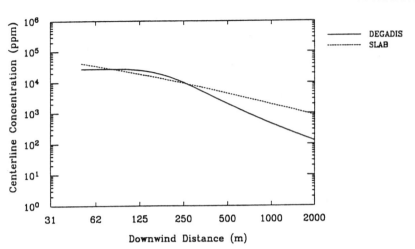

Figure D-3. Predicted cloud ground-level centerline concentrations as a function of downwind distance for DEGADIS and SLAB for Scenario 4 (normal butane jet) base run.

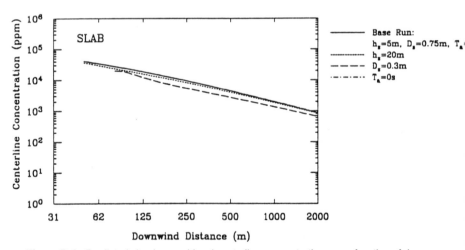

Figure D-4. Predicted cloud ground-level centerline concentrations as a function of downwind distance for SLAB for Scenario 4 (normal butane jet) for base run and three sensitivity runs.

effect on the SLAB predictions (increase of 40%). The change in averaging time has little effect on the touchdown distance, since the jet modules of SLAB and DEGADIS are not corrected for averaging time.

A hard-copy listing of DEGADIS output for the base run of Scenario 4 is given in Table D-4, and full listings for both models are on the diskette in the pocket inside the back cover.

TABLE D-3

Predicted Distance Where Plume Bottom Strikes Ground for Scenario 4 (Normal Butane Jet) for Base Run and Three Sensitivity Runs, for DEGADIS and SLAB

	Downwind Distance Where Plume Bottom Strikes Ground	
Run	**DEGADIS**	**SLAB**
Base Run (h_s = 5 m, D_s = 0.75 m, T_a = 600 s)	19 m	10 m
Sensitivity Run 1 (h_s = 20 m)	75 m	14 m
Sensitivity Run 2 (D_s = 0.3 m)	205 m	70 m
Sensitivity Run 3 (T_a = 0 s)	19 m	10 m

TABLE D-4   Output File for DEGADIS for Scenario 4 (Normal Butane Jet) Base Run

```
*********************                JETPLU/DEGADIS v2.1                *********************

16-AUG-1995 10:24:32.25

TRIAL AND EXPERIMENT: S4A        Scenario 4
CHEMICAL: Normal Butane
STEADY-STATE SIMULATION IN ISOTHERMAL MODE


Ambient Meteorological Conditions...                  Contaminant Properties...

Ambient windspeed at reference height:   2.0000   m/s       Contaminant molecular weight:   58.120
              Reference height:   10.000   m              Initial temperature:   303.00       K
              Surface roughness:   .10000   m           Upper level of interest:   1.00000E-02
                                                        Lower level of interest:   1.00000E-07
    Pasquill stability class:              F               Average heat capacity:   1715.0      J/kg K

    Monin-Obukhov length:        14.257    m

    Friction velocity:     8.84760E-02  m/s       NDEN flag:   0
    Ambient temperature:   303.00       K
    Ambient pressure:      1.0000       atm       ISOFL flag:   0
    Ambient humidity:      1.37707E-02
    Relative humidity:     50.000       %          Release Properties...

    Specified averaging time:    600.00    s                  Release rate:   15.000     kg/s
    DELTAy:   6.74000E-02                            Discharge elevation:   5.0000      m
    BETAy:    .90000                                 Discharge diameter:    .75000      m
    DELTAz:   1.12200E-02
    BETAz:    1.4024                        Model Parameters...
    GAMMAz:  -5.40000E-02
                                                             ALFA1:   2.80000E-02
                                                             ALFA2:   .37000
                                                             DISTMX:  6.0000       m
```

Downwind Distance (m)	Centerline Elevation (m)	Centerline Mole Fraction	Centerline Concentration (kg/m3)	Density (kg/m3)	Temperature (K)	Sigma y (m)	Sigma z (m)	At z = .000 m: Mole Fraction	Width to mol%: 1.00E-05 (m)	1.00E+00 (m)	Maximum Mole Fraction	Elevation for Max Mol Frac (m)
5.935E-02	10.1	1.00	2.34	2.34	303.	.413	.413	.000			1.00	10.1
3.14	15.3	.475	1.11	1.72	303.	1.20	1.19	.000			.475	15.3
6.61	14.0	.241	.564	1.44	303.	1.34	1.29	.000			.241	14.0
9.86	11.7	.142	.332	1.32	303.	1.65	1.58	.000			.142	11.7
13.8	8.95	8.766E-02	.205	1.26	303.	2.07	1.97	.000			8.766E-02	8.95
17.3	6.46	6.244E-02	.146	1.23	303.	2.45	2.32	.000			6.244E-02	6.46
20.3	4.13	4.837E-02	.113	1.21	303.	2.80	2.65	2.841E-02	6.01	4.04	4.839E-02	4.06
24.6	.549	6.721E-02	.157	1.23	303.	3.30	3.13	6.822E-02	7.10	6.47	6.822E-02	.000
25.3	.000	6.642E-02	.155	1.23	303.	3.37	3.20	6.561E-02	7.25	6.54	6.642E-02	.000

0 ******************   U O A _ D E G A D I S   M O D E L   O U T P U T   - -   V E R S I O N   2.1   *********************

0   TITLE BLOCK

TRIAL AND EXPERIMENT: S4A   Scenario 4
CHEMICAL: Normal Butane
STEADY-STATE SIMULATION IN ISOTHERMAL MODE

```
      Wind velocity at reference height        2.00   m/s
      Reference height                        10.00   m
0     Surface roughness length                 .100   m
0     Pasquill stability class                    F
0     Monin-Obukhov length                     17.5   m
      Gaussian distribution constants
            Specified averaging time         600.00   s
                              Deltay          .06740
                              Betay           .90000
                              Alpha           .44905
0     Wind velocity power law constant         .09593  m/s
      Friction velocity                      303.00   K
0     Ambient Temperature                     1.000    atm
0     Ambient Pressure                     1.377E-02   kg/kg BDA
      Ambient Absolute Humidity               50.00    %
      Ambient Relative Humidity
```

TABLE D-4 (Continued)

Adiabatic Mixing:

Mole fraction	CONCENTRATION OF C kg/m**3	GAS DENSITY kg/m**3	Enthalpy J/kg	Temperature K
.00000	.00000	1.15534	.00000	303.00
.06594	.15416	1.23332	.00000	303.00
.14144	.33065	1.32258	.00000	303.00
.22871	.53467	1.42577	.00000	303.00
.33075	.77322	1.54643	.00000	303.00
.45166	1.05587	1.68940	.00000	303.00
.59720	1.39612	1.86149	.00000	303.00
.77576	1.81355	2.07263	.00000	303.00
1.00000	2.33778	2.33778	.00000	303.00

0   Specified Gas Properties:
   Molecular weight: 58.120
   Release temperature: 303.00 K
   Density at release temperature and ambient pressure: 2.3378 kg/m**3
   Average heat capacity: 1715.0 J/kg K
   Upper mole fraction contour: 1.00000E-02
   Lower mole fraction contour: 1.00000E-07
   Height for isopleths: .00000 m

Source input data points

   Initial (pure contaminant) mass in cloud: .00000 kg

Time s	Contaminant Mass Rate kg/s	Source Radius m	Contaminant Mass Fraction kg contam/kg mix	Temperature K	Enthalpy J/kg
.00000	15.000	7.2535	.12585	303.00	.00000
60230.	15.000	7.2535	.12585	303.00	.00000
60231.	.00000	.00000	.12585	303.00	.00000
60232.	.00000	.00000	.12585	303.00	.00000

0   Calculation procedure for ALPHA: 1
0   Entrainment prescription for PHI: 3
0   Layer thickness ratio used for average depth: .590
0   Air entrainment coefficient used: 2.1500

   Gravity slumping velocity coefficient used: 1.150
   NON isothermal calculation
   Heat transfer not included
   Water transfer not included

Time sec	Gas Radius m	Height m	Qstar kg/m**2/s	SZ(x=L/2.) m	Mole frac C	Density kg/m**3	Temperature K	Rich No.
.000000	7.25353	1.100000E-05	3.587351E-03	.768550	6.642361E-02	1.23388	303.000	.756144
1.23328	7.86933	.632329	3.374092E-03	.820282	6.167345E-02	1.22827	303.000	.756144
3.12102	9.49003	1.28955	2.935973E-03	.949484	5.235679E-02	1.21725	303.000	.756144
5.00876	11.3413	1.66718	2.578478E-03	1.08627	4.521487E-02	1.20881	303.000	.756144
8.84403	15.1737	2.01863	2.097339E-03	1.34048	3.628135E-02	1.19824	303.000	.756144
12.6793	18.8718	2.13489	1.805204E-03	1.55688	3.126667E-02	1.19231	303.000	.756144
19.8857	25.3615	2.13958	1.483558E-03	1.88459	2.618230E-02	1.18630	303.000	.756144
27.0921	31.3420	2.04967	1.295982E-03	2.14060	2.350137E-02	1.18313	303.000	.756144
40.6594	41.5602	1.81606	1.091326E-03	2.49653	2.099911E-02	1.18017	303.000	.756144
54.2267	50.7489	1.57180	9.732102E-04	2.74116	1.996987E-02	1.17896	303.000	.756144
83.5053	67.9950	1.09071	8.359113E-04	3.01908	1.998047E-02	1.17897	303.000	.756144
112.784	77.3701	.801607	7.942877E-04	3.00982	2.169967E-02	1.18100	303.000	.756144
146.195	75.2971	.778454	8.220001E-04	2.83514	2.383296E-02	1.18352	303.000	.756144
198.890	74.2041	.765735	8.410889E-04	2.70828	2.568061E-02	1.18571	303.000	.756144
218.174	75.8058	.732008	8.332513E-04	2.70647	2.601566E-02	1.18610	303.000	.756144
288.504	74.8550	.721989	8.521177E-04	2.64369	2.717982E-02	1.18748	303.000	.756144

0Source strength [kg/s] : 15.000   Equivalent Primary source radius [m] : 7.2535
Equivalent Primary source length [m] : 14.507   Equivalent Primary source half-width [m] : 5.6969

Secondary source concentration [kg/m**3] : 6.35406E-02   Secondary source SZ [m] : 2.6437

Contaminant flux rate: 8.52118E-04

TABLE D-4 (Continued)

Secondary source mass fractions... contaminant: 5.350871E-02   air: .93363
                      Enthalpy: .00000   Density: 1.1875

Secondary source length [m] : 149.71   Secondary source half-width [m] : 58.791

0 Distance (m)	Mole Fraction	Concentration (kg/m**3)	Density (kg/m**3)	Temperature (K)	Half Width (m)	Sz (m)	Sy (m)	Width at z= 1.000E-05mole% (m)	Width at z= 1.00 mole% (m)	.00 m to: 1.00 mole% (m)
100.	2.718E-02	6.354E-02	1.19	303.	58.8	2.64	.000	58.8	58.8	
105.	2.683E-02	6.272E-02	1.19	303.	59.3	2.52	5.08	77.3	64.3	
132.	2.396E-02	5.602E-02	1.18	303.	80.0	2.10	14.9	132.	94.0	
172.	1.857E-02	4.342E-02	1.18	303.	110.	1.97	24.8	196.	130.	
212.	1.350E-02	3.157E-02	1.17	303.	136.	2.10	33.4	251.	154.	
252.	9.655E-03	2.257E-02	1.17	303.	157.	2.37	41.4	298.		
292.	6.994E-03	1.635E-02	1.16	303.	175.	2.73	48.9	338.		
332.	5.196E-03	1.215E-02	1.16	303.	190.	3.14	55.9	374.		
372.	3.971E-03	9.284E-03	1.16	303.	202.	3.59	62.6	406.		
412.	3.117E-03	7.286E-03	1.16	303.	213.	4.07	69.0	435.		
452.	2.505E-03	5.857E-03	1.16	303.	222.	4.57	75.1	461.		
492.	2.056E-03	4.807E-03	1.16	303.	230.	5.08	81.0	486.		
532.	1.718E-03	4.017E-03	1.16	303.	237.	5.59	86.8	508.		
572.	1.458E-03	3.409E-03	1.16	303.	243.	6.11	92.3	529.		
612.	1.254E-03	2.932E-03	1.16	303.	249.	6.64	97.7	549.		
652.	1.092E-03	2.552E-03	1.16	303.	254.	7.16	103.	568.		
692.	9.598E-04	2.244E-03	1.16	303.	259.	7.69	108.	586.		
732.	8.514E-04	1.990E-03	1.16	303.	263.	8.22	113.	603.		
772.	7.611E-04	1.780E-03	1.16	303.	267.	8.74	118.	619.		
812.	6.855E-04	1.602E-03	1.16	303.	270.	9.27	123.	634.		
852.	6.211E-04	1.452E-03	1.16	303.	273.	9.79	127.	649.		
892.	5.659E-04	1.323E-03	1.16	303.	276.	10.3	132.	664.		
932.	5.181E-04	1.212E-03	1.16	303.	279.	10.8	136.	678.		
972.	4.768E-04	1.115E-03	1.16	303.	282.	11.3	141.	691.		

1.012E+03	4.405E-04	1.030E-03	1.16	303.	284.	11.9	145.	704.
1.052E+03	4.085E-04	9.549E-04	1.16	303.	286.	12.4	149.	717.
1.092E+03	3.801E-04	8.885E-04	1.16	303.	288.	12.9	153.	729.
1.132E+03	3.548E-04	8.294E-04	1.16	303.	290.	13.4	158.	741.
1.172E+03	3.322E-04	7.765E-04	1.16	303.	292.	13.9	162.	752.
1.212E+03	3.118E-04	7.290E-04	1.16	303.	294.	14.4	166.	764.
1.252E+03	2.935E-04	6.860E-04	1.16	303.	295.	14.9	170.	775.
1.292E+03	2.768E-04	6.471E-04	1.16	303.	297.	15.4	173.	785.
1.332E+03	2.617E-04	6.117E-04	1.16	303.	298.	15.9	177.	796.
1.412E+03	2.352E-04	5.499E-04	1.16	303.	301.	16.8	185.	816.
1.452E+03	2.236E-04	5.228E-04	1.16	303.	302.	17.3	189.	826.
1.532E+03	2.031E-04	4.748E-04	1.16	303.	304.	18.3	196.	845.
1.572E+03	1.940E-04	4.535E-04	1.16	303.	305.	18.7	200.	854.
1.652E+03	1.777E-04	4.154E-04	1.16	303.	307.	19.7	207.	872.
1.692E+03	1.704E-04	3.983E-04	1.16	303.	308.	20.2	210.	881.
1.772E+03	1.572E-04	3.674E-04	1.16	303.	310.	21.1	217.	898.
1.812E+03	1.512E-04	3.535E-04	1.16	303.	310.	21.5	220.	907.
1.892E+03	1.404E-04	3.281E-04	1.16	303.	312.	22.5	227.	923.
1.932E+03	1.354E-04	3.166E-04	1.16	303.	312.	22.9	230.	931.
2.012E+03	1.264E-04	2.954E-04	1.16	303.	313.	23.8	237.	946.
2.052E+03	1.222E-04	2.857E-04	1.16	303.	314.	24.3	240.	954.
...								
1.527E+05	9.804E-08	2.292E-07	1.16	303.	.000	795.	4.717E+03	

For the UFL of   1.0000   mole percent, and the LFL of   1.00000E-05 mole percent:

The mass of contaminant between the UFL and LFL is: 2.33288E+05 kg.
The mass of contaminant above the LFL is: 2.34791E+05 kg.

APPENDIX E

Scenario 5: Release of SO_3 within a Building and Subsequent Dispersion of H_2SO_4 from a Vent on the Building

Scenario 5 involves the release of 408 kg of SO_3 due to a runaway reaction inside a square building of volume 9000 m^3, containing a mass, 10,800 kg, of air. The height of the building is 18 m. The release in the building extends over a time period of about five minutes. The building is vented through a stack located on the roof top, where the plume is emitted at a constant speed of 6.37 m/s and a volume flow rate of 5 m^3 s^{-1}. Therefore, the time scale for an air exchange in the building is (9000 m^3)/(5 m^3/s) = 1800 seconds or 30 minutes. The stack height is assumed to be 20 m above the ground and 2 m above the building top, and the stack diameter is assumed to be 1 m. The ambient atmosphere is assumed to be slightly unstable (class C) with a wind speed of 2 m s^{-1} at a reference height of 10 m. A schematic diagram of this scenario is given in Figure E-1.

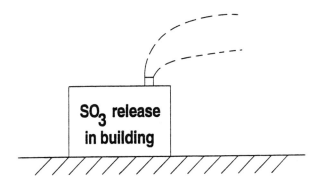

Figure E-1. Schematic diagram of Scenario 5, H_2SO_4 release from a building vent.

Scenario 5 is different from the other six scenarios in that the plume is less dense than air when it enters the atmosphere, due to heat released by chemical reactions of SO_3 with water vapor, and the effects of building downwash need to be investigated, since the vent is not far above the top of the building. Since the plume is not dense, the standard EPA models ISC3 and INPUFF are chosen for application to Scenario 5.

Inside the building, where the air is assumed to be well-mixed, the 408 kg (5.1 kmole) of SO_3 that is released is assumed to react completely with 92 kg (5.1 kmole) of water vapor to form 500 kg of H_2SO_4 mist. The mass of water vapor required for this reaction is determined by multiplying 5.1 kmole by the molecular weight of H_2O (18 kg/kmole). There is enough water vapor inside the building, assuming 50% relative humidity, to fully react with the SO_3. Assuming that the 500 kg of H_2SO_4 is formed inside the building within a few minutes and is well-mixed throughout the building, the concentration, C_o, initially in the building is then equal to 500 kg/9000 $m^3 = 0.0555$ kg m^{-3}. This concentration represents an approximate 6% mixture of H_2SO_4 in air. Given this C_o, the initial mass release rate, Q_o, of H_2SO_4 from the vent is 0.0555 kg $m^{-3} \cdot 5$ m^3 $s^{-1} = 0.278$ kg s^{-1}.

Although we have no detailed knowledge of the air circulation patterns in the building, we do know the magnitude of the constant ventilation rate. In this case, it is common practice to assume that the air is continually well-mixed and, consequently, that the concentration, C, in the building decreases exponentially with a time scale equal to the air exchange time (30 minutes in the case of Scenario 5). This assumption is also often used by chemical engineers in their analysis of well-mixed reactors. Therefore, in the building, it is assumed that the concentration varies with time according to the exponential formula:

$$C = C_o e^{-t/30\text{min}} \tag{E-1}$$

where time, t, is in minutes. The gas release rate of H_2SO_4 from the vent also follows an exponential formula:

$$Q(t) = Q_o e^{-t/30 \text{ min}} \tag{E-2}$$

The average emission rate, Q_a (t_2, t_1), between times t_1 and t_2 is given by simple integration of equation (E-2):

$$Q_a(t_2, t_1) = Q_o \left(e^{-t_1/30 \text{ min}} - e^{-t_2/30 \text{ min}}\right) \frac{30 \text{ min}}{t_2 - t_1} \tag{E-3}$$

The solution to equation (E-3) is tabulated below in ten-minute increments for times after release ranging from 0 to 60 min.

t_1 (min)	t_2 (min)	Q_a (t_2, t_1) (g s^{-1})
0	10	236
10	20	169
20	30	121
30	40	87.0
40	50	62.3
50	60	44.7

The fraction of initial mass remaining after time t_2 is given by exp $(-t_2/30$ min$)$. Therefore, 37% of the mass remains after 30 min, and 13.5% of the mass remains after 60 min.

The reaction of H_2O and SO_3 results in a release of heat (41.5 kcal mole^{-1}), which increases the temperature of the mixture in the building and therefore causes the plume coming out of the vent stack to be buoyant (less dense than air). The total heat released by the reaction within the building equals the number of moles of SO_3 released (5100 moles) times 41.5 kcal mole^{-1}, or 2.12×10^5 kcal. This heat release would increase the temperature of the air–H_2SO_4 mixture in the building by an amount, ΔT, given by the formula

$$\Delta T = \frac{2.12 \times 10^5 \text{ kcal}}{10,800 \text{ kg} \cdot c_p} = 78 \text{ K} \qquad \text{(E-4)}$$

where the specific heat for air, c_p, equals 0.25 kcal kg^{-1} K^{-1}. There is assumed to be 10,800 kg of air in the building. The effects of an overpressure due to thermal expansion on the flow through the vent stack are assumed to be minimal.

The trajectory of the plume from the vent can be calculated using plume rise formulas summarized in Section 5 and described in detail by Briggs (1984), Turner (1994), and EPA (1995) once the initial momentum and buoyancy fluxes are determined. The initial momentum flux (maintained throughout the incident by a fan in the vent), M_0, is defined in the paragraph after equation (5-5) by the following expression

$$M_0 = w_0^2 R_0^2 \qquad \text{(E-5)}$$

where w_0 is the initial jet speed (6.37 m s^{-1}) and R_0 is the stack radius (0.5 m). Thus, $M_0 = 10.1$ m^4 s^{-2}, which is constant with time.

The initial buoyancy flux, F, is given by:

$$F = w_0 R_0^2 \, g \, \frac{T_p - T_a}{T_p} \qquad \text{(E-6)}$$

where g is acceleration due to gravity (9.8 m s^{-2}), T_a is the ambient temperature (300 K), and T_p is the initial plume temperature (300 K + 78 K = 378 K).

Thus, $F_0 = 3.2$ m^4 s^{-3}, which is likely to decay exponentially with time in the same manner as the mass source term, since the excess temperature in the vent plume decays exponentially with time:

$$F_0(t) = (3.2 \text{ m}^4 \text{ s}^{-3}) \, e^{-t/30 \text{ min}}$$

$$(E-7)$$

Therefore, the following initial plume conditions can be assumed for the H$_2$SO$_4$ jet emitted from the vent:

Time Increment (min)	Mass Emission Rate, Q (kg s^{-1})	Momentum Flux, M_0 (m^4 s^{-2})	Temperature Excess, $T_p - T_a$ (K)	Buoyancy Flux, F_0 (m^4 s^{-3})
0	0.278	10.1	78	3.2
0–10	0.236	10.1	66.3	2.8
10–20	0.169	10.1	47.5	1.95
20–30	0.121	10.1	34.1	1.39
30–40	0.087	10.1	24.4	1.00
40–50	0.0623	10.1	17.5	0.71
50–60	0.0447	10.1	12.5	0.52

There is a question whether the plume is affected by distortions to the air flow caused by the building. The plume rise, Δh, needs to be estimated and compared to the building height. The following standard formula from the ISC manual (EPA, 1995) can be used to calculate the buoyant plume final rise:

$$\Delta h = 21.4 \frac{F_0^{3/4}}{u}$$

$$(E-8)$$

where u is the wind speed at the stack height. The following table lists the values for plume rise for selected values of F_0 and u.

	Plume Rise, Δh		
Initial Flux	$u = 1$ m s^{-1}	$u = 2$ m s^{-1}	$u = 3$ m s^{-1}
$F_0 = 3$ m^4 s^{-3}	49 m	24 m	16 m
$F_0 = 2$ m^4 s^{-3}	36 m	18 m	12 m
$F_0 = 1$ m^4 s^{-3}	21 m	10 m	7 m

The downwind distance to final rise, x_f, is $49 \, F_o^{5/8}$, according to the ISC manual (EPA, 1995). Values of calculated distances, x_f, are given below:

Buoyancy Flux F_o ($m^4 \, s^{-3}$)	Downwind Distance x_f (m)
3	90 m
2	76 m
1	49 m

Therefore the plume final rise is reached at $x \pm 100$ m for wind speeds less than 3 m s⁻¹.

Since the vent stack is located on the roof top, building downwash could become important for the scenario when $h_s + \Delta h < 2.5 \, h_b$, where h_s is the stack height and h_b is the building height. For $h_s = 20$ m and $h_b = 18$ m, the data in the above table suggest that building downwash is expected to be important for most cases. ISC will take this into account, but INPUFF ignores the presence of the building.

A summary of the input conditions for the base run and the three sensitivity runs for Scenario 5 is given in Table E-1. These inputs are sufficient to run INPUFF and ISC3.

The EPA's INPUFF and ISC3 dispersion models have been applied, since this is a buoyant plume from a vent stack on an industrial building, and the mass release rate can be assumed to be constant over ten-minute blocks of time. INPUFF was chosen because it can treat the time-variable nature of the release by tracking the transport speed and dilution rate of discrete emissions of puffs. ISC3 was chosen because it is the standard model applied by the EPA to industrial sources; however, ISC3 assumes a continuous steady-state plume. It is not necessary to check the source Richardson number, Ri_o, for this scenario, since the plume is positively buoyant. INPUFF requires the user to specify the initial lateral and vertical dispersion coefficients (σ_{yo} and σ_{zo}). If not known, it is recommended that σ_{yo} and σ_{zo} are assumed equal and given by:

$$\sigma_{yo} = \sigma_{zo} = R_{\text{eff}} = \left(\frac{Q_o}{\rho \pi u_{st}} \right)^{1/2} \qquad \text{(E-9)}$$

where u_{st} is the wind speed at the stack height, h_s.

This formula gives $\sigma_{yo} = \sigma_{zo} = 0.4$ m. The resulting predictions of cloud centerline concentrations (at ground-level) are listed in Table E-2 and are plotted in Figures E-2 and E-3 for the base run and for the three sensitivity

TABLE E-1
Input Conditions for Scenario 4, H_2SO_4 Release from Building
(Data Taken from Table 9-2, Which Covers All Seven Scenarios)

SULFURIC ACID PROPERTIES	
molecular weight	$M = 98.04$ kg kmole^{-1}
SOURCE CONFIGURATION	
stack vent height	$h_s = 20$ m
building height	$h_b = 18$ m
building width (square)	$W_b = 22.36$ m
vent flow rate	$V_0 = 5$ m^3 s^{-1}
mass emission rate	$Q_0 = 0.272$ kg s^{-1}, exponentially decreasing with 30-min time scale.
plume buoyancy flux	$F_0 = 3.2$ m^4 s^{-3}, exponentially decreasing with 30-min time scale.
METEOROLOGICAL CONDITIONS	
ambient temperature	$T_a = 300$ K
wind speed	$u = 2$ m s^{-1}
stability class	C
relative humidity	RH = 50%
Monin-Obukhov length	$L = -74.2$ m
friction velocity	$u_* = 0.303$ m s^{-1}
surface roughness	$z_0 = 0.5$ m (industrial area)
SENSITIVITY RUNS	
Sensitivity Run 1	Change stability to class F and decrease wind speed to 2 m s^{-1}
Sensitivity Run 2	Increase building height from 18 m to 36 m; increase stack vent height from 20 m to 38 m
Sensitivity Run 3	Increase stack vent height from 20 m to 40 m
CONCENTRATIONS OF INTEREST	
averaging time $T_a = 600$ s	IDLH $C = 20$ ppm

runs. The ISC3 predictions near the source (i.e., at $x = 100$ or 200 m) are larger than the INPUFF predictions because, unlike INPUFF, ISC3 accounts for the possibility of downwash of the plume into the building wake. However, even so, the predicted concentrations do not exceed the IDLH concentration of 20 ppm in the downwind distance range from 100 m to 2000 m. Figure E-2 shows that, at larger downwind distances of about 500 m or greater, where the plume is expected to be well-mixed between the stack height and the ground, the INPUFF predictions exceed the ISC predictions by a factor of four or five. These differences are probably due to differences in the σ_y and σ_z formulas used by the two models.

TABLE E-2
Predicted Cloud Centerline Concentrations (at Ground-Level) as a Function of Downwind Distance for INPUFF and ISC for Four Model Input Conditions (Base Run plus Three Sensitivity Runs) for Scenario 5 (H_2SO_4 Release)

		SCENARIO 5	
		Centerline Concentration (ppm)	
		INPUFF	ISC
Base Run:	100 m	0.051	9.4
C/2	200 m	0.88	4.1
h_b=18m	500 m	1.8	.0.8
h_s=20m	1000 m	0.95	0.22
	1500 m	0.5	0.1
	2000 m	0.29	0.063
Sensitivity Run 1:	100 m	1.41E-12	7.6
F/2	200 m	6.24E-05	5.9
	500 m	0.061	3.2
	1000 m	0.39	1.4
	1500 m	0.52	0.79
	2000 m	0.53	0.54
Sensitivity Run 2:	100 m	4.77E-05	2.3
h_b=36m	200 m	4.60E-02	2.6
h_s=38m	500 m	0.83	0.67
	1000 m	0.73	0.19
	1500 m	0.44	0.092
	2000 m	0.26	0.055
Sensitivity Run 3:	100 m	2.16E-05	0.31
h_b=18m	200 m	0.031	1.5
h_s=40m	500 m	0.75	0.6
	1000 m	0.71	0.18
	1500 m	0.44	0.09
	2000 m	0.26	0.054

The results of the three sensitivity runs are plotted for INPUFF in Figure E-3. When the stability class is changed from class C (slightly unstable) to class F (stable), the predicted vertical diffusion coefficient, σ_z, is greatly reduced, causing the plume to not come down to the ground until much farther downwind (say, 500 m instead of 100 m). However, once the stable (class F) plume thoroughly mixes to the ground, at downwind distances

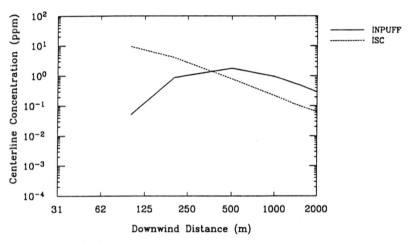

Figure E-2. Predicted cloud centerline concentrations (at ground level) as a function of downwind distance for INPUFF and ISC for the base run for Scenario 5 (H_2SO_4 release).

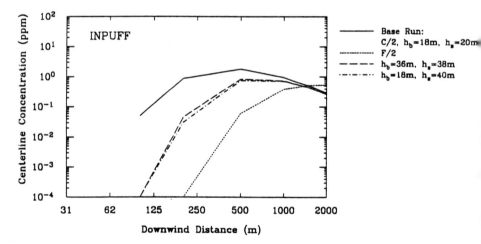

Figure E-3. Predicted cloud centerline concentrations (at ground level) as a function of downwind distance for INPUFF for the base run and the three sensitivity runs for Scenario 5 (H_2SO_4 release)

beyond about 1500 m, its predicted concentrations are greater than the predicted concentrations for the unstable (class C) plume. The predicted concentration resulting from changes in building height and stack height are seen to be within a few percent in Figure E-3. In both cases, the stack height approximately doubles (to about 40 m), and the predicted elevated plume does not disperse to the ground as quickly. It should be recalled that INPUFF does not account for the effects of the building. It is seen, though, that since all other conditions remain the same, the predicted concentration for h_s = 40 m approaches the predicted concentration for h_s = 20 m at large downwind distances (beyond about 1000 m).

A hard-copy of the output file for INPUFF for the base run of Scenario 5 is given in Table E-3. This file as well as similar ISC files for all runs are included on the diskette in the pocket inside the back cover.

TABLE E-3 Output File for INPUFF for Scenario 5 Base Run

```
M O D E L   O P T I O N S         A "T" INDICATES THAT
                            THE OPTION HAS BEEN EXERCISED

USER SUPPLIED WIND FIELD              F
UNIT 22 OUTPUT OPTION                 F
PRINT PUFF INFORMATION               F
INTERMEDIATE CONCENTRATIONS          T

DISPERSION CALCULATED USING PASQUILL-GIFFORD (DISTANCE DEPENDENT) SIGMA CURVES,
WITH TRANSITION TO DRAXLER'S LONG RANGE TRANSPORT SIGMA-Y AT SYMAX =    5000.0 METERS.

         B E G I N   A N A L Y S I S   O F   S O U R C E   N U M B E R   1

S O U R C E   O P T I O N S         A "T" INDICATES THAT
                            THE OPTION HAS BEEN EXERCISED

STACK DOWNWASH                       T
BUOYANCY INDUCED DISPERSION          T
DEPOSITION AND SETTLING              F
USER PLUME RISE                      F
PERFORM PUFF COMBINATIONS            T

I N P U T   P A R A M E T E R S

SOURCE UPDATE INTERVAL =      600 SECONDS.     (-1 INDICATES NO UPDATE)
START CONCENTRATION CALCULATIONS AT TIME =      0 SECONDS.
ANEMOMETER HEIGHT =   10.0 METERS.

* * *   I N F O R M A T I O N   F O R   S O U R C E   N U M B E R   1   * * *
```

SOURCE STRENGTH (G/SEC)	STACK HEIGHT (M)	STACK TEMP. (DEG-K)	STACK GAS VELOCITY (M/SEC)	STACK DIAMETER (M)	VOLUME FLOW (M**3/SEC)	COORD. AT TIME 0 SECONDS EAST (KM)	NORTH (KM)
.236E+03	20.00	366.300	6.370	1.000	4.422	0.000	0.000

230

SOURCE SPEED (M/SEC)	SOURCE DIRECTION (DEG)	PLUME HEIGHT (M)	INITIAL SIGMAS (R) (M)	(Z) (M)	DEPOSITION VELOCITY (CM/SEC)	SETTLING VELOCITY (CM/SEC)
0.000	0.0	41.79	0.1	0.1	0.00	0.00

* * *   S O U R C E   U P D A T E   T A B L E   * * *

TIME (SEC)	SOURCE STRENGTH (G/SEC)	STACK HEIGHT (M)	STACK TEMP. (DEG-K)	STACK GAS VELOCITY (M/SEC)	STACK DIAMETER (M)	VOLUME FLOW (M**3/SEC)	SOURCE SPEED (M/SEC)	SOURCE DIRECTION (DEG)	INITIAL SIGMAS (R) (M)	(Z) (M)
0	.236E+03	20.00	366.300	6.370	1.000	4.422	0.000	0.0	0.1	0.1
600	.169E+03	20.00	347.500	6.370	1.000	3.169	0.000	0.0	0.1	0.1
1200	.121E+03	20.00	334.100	6.370	1.000	2.270	0.000	0.0	0.1	0.1
1800	.870E+02	20.00	324.400	6.370	1.000	1.627	0.000	0.0	0.1	0.1
2400	.623E+02	20.00	317.500	6.370	1.000	1.166	0.000	0.0	0.0	0.0
3000	.447E+02	20.00	312.500	6.370	1.000	0.835	0.000	0.0	0.0	0.0

* * *   M E T E O R O L O G Y   * * *

WIND DIR. (DEG)	WIND SPD. (M/SEC)	MIXING HGT. (M)	PROF.EP (DIMEN)	STABILITY (CLASS)	U PLUME (M/SEC)	TEMP (K)	SIGMA TH. (RAD.)	SIGMA PH. (RAD.)
270.0	2.000	5000.	0.100	3	2.307	300.0	******	******

	PUFF RELEASE RATE (SEC)	SOURCE RECEPTOR DISTANCE (KM)	PUFF COMB. CRITERION (SIGMAS)
	10.000	0.10	1.000

SIMULATION PERIOD START (SEC) STOP (SEC)	SIMULATION TIME (SEC)
0 3600	3600

TABLE E-3 (Continued)

600 SEC AVG. CONCENTRATION AT RECEPTORS DURING INTERMEDIATE PERIOD   0 TO   600 SECONDS

RECEPTORS			
X (KM)	Y (KM)	Z (M)	CONCENTRATION (G/M**3)
0.100	0.000	0.000	4.926E-05
0.200	0.000	0.000	2.088E-03
0.500	0.000	0.000	5.022E-03
1.000	0.000	0.000	1.123E-03
1.500	0.000	0.000	2.010E-05
2.000	0.000	0.000	1.183E-11

600 SEC AVG. CONCENTRATION AT RECEPTORS DURING INTERMEDIATE PERIOD   600 TO   1200 SECONDS

RECEPTORS			
X (KM)	Y (KM)	Z (M)	CONCENTRATION (G/M**3)
0.100	0.000	0.000	7.597E-05
0.200	0.000	0.000	2.906E-03
0.500	0.000	0.000	7.169E-03
1.000	0.000	0.000	3.780E-03
1.500	0.000	0.000	1.980E-03
2.000	0.000	0.000	7.507E-04

600 SEC AVG. CONCENTRATION AT RECEPTORS DURING INTERMEDIATE PERIOD 1200 TO   1800 SECONDS

RECEPTORS			
X (KM)	Y (KM)	Z (M)	CONCENTRATION (G/M**3)
0.100	0.000	0.000	1.084E-04
0.200	0.000	0.000	3.304E-03
0.500	0.000	0.000	5.931E-03
1.000	0.000	0.000	2.847E-03
1.500	0.000	0.000	1.691E-03
2.000	0.000	0.000	1.158E-03

600 SEC AVG. CONCENTRATION AT RECEPTORS DURING INTERMEDIATE PERIOD 1800 TO 2400 SECONDS

| RECEPTORS | | | |
X (KM)	Y (KM)	Z (M)	CONCENTRATION (G/M**3)
0.100	0.000	0.000	1.530E-04
0.200	0.000	0.000	3.487E-03
0.500	0.000	0.000	4.752E-03
1.000	0.000	0.000	2.147E-03
1.500	0.000	0.000	1.239E-03
2.000	0.000	0.000	8.564E-04

600 SEC AVG. CONCENTRATION AT RECEPTORS DURING INTERMEDIATE PERIOD 2400 TO 3000 SECONDS

| RECEPTORS | | | |
X (KM)	Y (KM)	Z (M)	CONCENTRATION (G/M**3)
0.100	0.000	0.000	2.033E-04
0.200	0.000	0.000	3.430E-03
0.500	0.000	0.000	3.696E-03
1.000	0.000	0.000	1.592E-03
1.500	0.000	0.000	9.137E-04
2.000	0.000	0.000	6.310E-04

600 SEC AVG. CONCENTRATION AT RECEPTORS DURING INTERMEDIATE PERIOD 3000 TO 3600 SECONDS

| RECEPTORS | | | |
X (KM)	Y (KM)	Z (M)	CONCENTRATION (G/M**3)
0.100	0.000	0.000	1.539E-04
0.200	0.000	0.000	2.596E-03
0.500	0.000	0.000	2.739E-03
1.000	0.000	0.000	1.165E-03
1.500	0.000	0.000	6.610E-04
2.000	0.000	0.000	4.582E-04

TABLE E-3 (Continued)

3600 SEC AVG. CONCENTRATION AT RECEPTORS FOR SIMULATION PERIOD   0 TO   3600 SECONDS
            DUE TO SOURCE NUMBER 1

RECEPTORS			
X (KM)	Y (KM)	Z (M)	CONCENTRATION (G/M**3)
0.100	0.000	0.000	1.240E-04
0.200	0.000	0.000	2.968E-03
0.500	0.000	0.000	4.885E-03
1.000	0.000	0.000	2.109E-03
1.500	0.000	0.000	1.084E-03
2.000	0.000	0.000	6.424E-04

* *

1.00 HR AVG. CONCENTRATION AT RECEPTORS FOR ALL SIMULATION PERIODS
            DUE TO SOURCE NUMBER 1

RECEPTORS			
X (KM)	Y (KM)	Z (M)	CONCENTRATION (G/M**3)
0.100	0.000	0.000	1.240E-04
0.200	0.000	0.000	2.968E-03
0.500	0.000	0.000	4.885E-03
1.000	0.000	0.000	2.109E-03
1.500	0.000	0.000	1.084E-03
2.000	0.000	0.000	6.424E-04

APPENDIX F

Scenario 6: Pressurized Horizontal Gas Jet Release of a Multicomponent Mixture

Scenario 6 is concerned with a multicomponent gas release from a horizontal nozzle on a pressurized tank. The nozzle has diameter 0.1 m and is located at a height of 1 m above the ground (see Figure F-1). A gaseous mixture of four components is assumed, with mole fractions of 60% ethane, 10% propane, 20% nitrogen, and 10% oxygen. The mixture is stored at 293 K and 4 atm. The ambient atmosphere is assumed to be neutral (class D) with a wind speed of 5 m s^{-1} at a reference height of 10 m. Tables 9-2 and F-1 list the source configuration, the meteorological data, and the site conditions for Scenario 6.

The calculation of the source term for multicomponent releases can be complicated, as discussed in Section 4. For the purposes of this example scenario, it is arbitrarily assumed that the source term is 5 kg s^{-1}. Since the release is a gas, the only mixture properties that are needed are the effective molecular weight (31.25 kg kmole^{-1}) and the effective specific heat for vapor (1513 J kg^{-1} K^{-1}), which represent weighted averages over all four components.

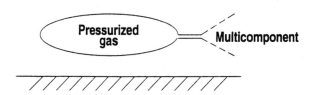

Figure F-1. Schematic diagram of Scenario 6, the pressurized horizontal jet release of a multicomponent mixture.

TABLE F-1
Input Data Archive for Scenario 6, a Pressurized Horizontal Gas Jet Release
of a Multicomponent Mixture (Data Taken from Table 9-2, Which Covers All
Seven Scenarios)

GAS PROPERTIES	
components by mole: 60% ethane, 10 % propane, 20% nitrogen, 10 % oxygen	
molecular weight	$M = 31.25$ kg kmole^{-1} (weighted)
specific heat at constant pressure	$c_p = 1513$ J kg^{-1} K^{-1} (weighted)

SOURCE CONFIGURATION	
storage and release temperature	$T_s = 293$ K
storage pressure	$p_s = 4$ atm
nozzle diameter	$D_p = 0.1$ m
nozzle orientation	horizontal, pointing downwind
nozzle height above ground	$h_s = 1$ m
source strength	$Q = 5$ kg s^{-1}

METEOROLOGICAL CONDITIONS	
temperature	$T_a = 303$ K at height of 10 m
wind speed	$u = 5$ m s^{-1} at height of 10 m
stability class	D
Monin-Obukhov length	$L = 9999$ m (neutral)
friction velocity	$u_* = 0.51$ m s^{-1}
surface roughness	$z_0 = 0.2$ m (commercial area)

SENSITIVITY RUNS	
RUN 1	Change stability class to F (stable) and wind speed to 2 ms^{-1}
RUN 2	Increase source term by factor of 4.
RUN 3	Change averaging time to 0 s.

CONCENTRATIONS OF INTEREST			
averaging time	$T_a = 600$ s	IDLH	$C = 50,000$ ppm
UEL	$C = 125,000$ ppm	LEL	$C = 30,000$ ppm

It is advisable to first calculate the initial Richardson number, Ri_o, to determine if it is large enough to justify application of dense gas models. Since this is an elevated dense gas jet, equations (5-3a) and (5-4a) are valid:

For maximum ground-level concentration at plume touchdown:

$$Ri_o = \frac{g(\rho_{po} - \rho_a)V_{co}}{\rho_a u_*^3 h_s} \qquad \text{(F-1)}$$

For jet trajectory very near source:

$$Ri_o = \frac{g(\rho_{po} - \rho_a)V_{co}}{\rho_a u_*^3 D_p} \tag{F-2}$$

The following variables are known from Table F-1 or are calculated from known parameters:

ρ_{po} = 1.30 kg m^{-3}

ρ_a = 1.16 kg m^{-3}

h_s = 1 m

D_p = 0.1 m

u_* = 0.511 m s^{-1}

V_{co} = $Q/(\pi\rho_{po})$ = 1.223 m^3 s^{-1}

The resulting values of Ri_o equal 10.8 for maximum ground-level concentration, and 108 for plume behavior near the source. Since the critical Ri_o is 50, it is suggested that dense gas models are appropriate for simulating the jet trajectory near the source, but that the plume is not very dense at distances beyond the point of plume touchdown.

The ALOHA, HGSYSTEM, and SLAB vapor cloud dispersion models are applied to Scenario 6, given the initial conditions in Table F-1. To run SLAB or HGSYSTEM, the averaged or weighted molecular weight and specific heat for the mixture must be specified in the input file. However, since ALOHA retrieves the pollutant properties from a chemical database, the characteristics of the mixture of interest must be added to the database as a "pseudo-compound" using the built-in ALOHA chemical database manager. If the source term is directly specified by the user (i.e., the source term calculations are not carried out by ALOHA), then the user needs to specify the molecular weight, the normal boiling point, the gas density, the specific heat of the gas, and the vapor pressure for the new chemical in ALOHA's chemical database. The mixture molecular weight and specific heat were calculated by mole-weighting. The mixture gas density was derived from the mixture molecular weight at the given temperature and pressure using the ideal gas law. The values of the vapor pressure and the normal boiling point are not used in the calculations, although they should be consistent with the fact that the release is gaseous.

Comparisons of predicted cloud centerline concentrations as a function of downwind distance are given in Table F-2 and Figures F-2 and F-3. The SLAB and ALOHA predictions in Figure F-2 are within a factor of two at distances in the range from 100 m to 200 m, and the SLAB predictions are about 30% less than the ALOHA predictions at 2000 m. The HGSYSTEM

TABLE F-2
Predicted Cloud Centerline Concentration (ppm) as a Function of Downwind
Distance for ALOHA, HGSYSTEM and SLAB for the Base Run and Three
Sensitivity Runs for Scenario 6 (Multicomponent Gas Jet)

| | | SCENARIO 6 | | |
| | | Centerline Concentration (ppm) | | |
		ALOHA	HGSYSTEM	SLAB
Base Run:	100 m	9120	1311	4486
D/5	200 m	2980	583	1690
Q=5kg/s	500 m	574	156	362
T_a=600s	1000 m	158	51	104
	1500 m	74	27	50
	2000 m	43	17	31
Sensitivity Run 1:	100 m	19100	4949	7345
F/2	200 m	6700	2960	4750
	500 m	1570	1857	2621
	1000 m	528	997	1518
	1500 m	282	637	1018
	2000 m	182	450	747
Sensitivity Run 2:	100 m	22600	5898	5695
Q=20kg/s	200 m	8500	3306	3090
	500 m	2110	514	1086
	1000 m	607	120	392
	1500 m	287	53	200
	2000 m	167	30	124
Sensitivity Run 3:	100 m	9120	6346	5022
T_a=0s	200 m	2980	2776	2200
	500 m	574	725	590
	1000 m	158	233	196
	1500 m	74	124	99
	2000 m	43	79	62

predictions are close to the SLAB predictions at distances less than 100 m,
where HGSYSTEM uses the AEROPLUME algorithm. However, at distances beyond 100 m, after HGSYSTEM "transitions" to the PGPLUME
algorithm (with a relatively sharp factor of four drop in concentration seen
in Figure F-2), the HGSYSTEM predictions are less than the SLAB
predictions by a factor of two or three. The rate of decrease of concentration
with distance at distances beyond about 100 m is nearly the same in Figure
F-2 for the three models.

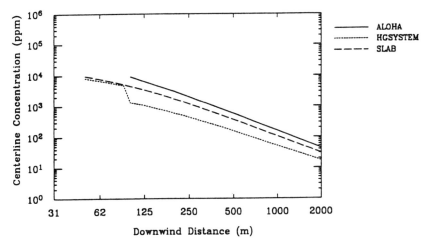

Figure F-2. Predicted cloud centerline concentrations as a function of downwind distance for ALOHA, HGSYSTEM, and SLAB for Scenario 6 (multicomponent gas jet) base run.

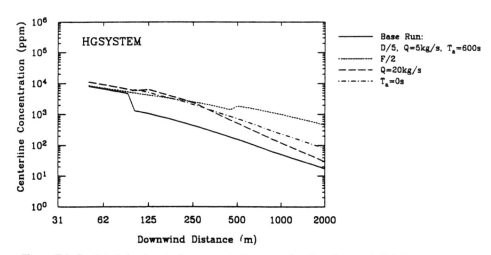

Figure F-3. Predicted cloud centerline concentrations as a function of downwind distance for HGSYSTEM for Scenario 6 for base run and three sensitivity runs.

Figure F-3 presents the results of the three sensitivity runs with HGSYSTEM. Transitions from one HGSYSTEM module to another are seen to be marked by discontinuities in the slopes of the curves. These discontinuities are artifacts of the coding and are not physically realistic. Depending on the input conditions, the near-field AEROPLUME module can make a transition to the HEGADAS module if the plume is denser than air at the transition point, or to the PGPLUME module if the plume is neutrally buoyant at the transition point. A transition to HEGADAS took place for Sensitivity Runs 1 and 2, while a transition to PGPLUME took place for Sensitivity Run 3. When stability class and wind speed are changed from D/5 to F/2, the predicted concentrations increase by a factor of 20 to 30 at downwind distances of 1000 to 2000 m. Part of this increase is due to the factor of 2.5 reduction in dilution by the wind, and the rest of the increase is due to the reduction in dispersion during stable conditions. When the source term is increased by a factor of four (from 5 kg s^{-1} to 20 kg s^{-1}), the predicted concentrations also increase, as expected, by a factor of two to ten, depending on downwind distance. A rather unexpected result is seen when averaging time, T_a, is decreased from 600 s to 0 s. Instead of seeing a factor of two increase in concentration, as typically observed and noted in Section 6, there is a factor of four to five increase in concentrations at distances beyond 100 m. This difference is due to the fact that PGPLUME assumes that both the lateral dispersion coefficient, σ_y, and the vertical dispersion coefficient, σ_z, vary with T_a according to $T_a^{1/5.}$ Therefore, the predicted concentration varies with T_a according to $T_a^{-2/5}$, rather than the $T_a^{-1/5}$ relation assumed by other models (Wilson, 1995).

The diskette in the pocket inside the back cover contains the model output files for all runs. An example of screen output for ALOHA for the base run of Scenario 6 is given as hard copy in Table F-3.

TABLE F-3
Screen Output for ALOHA for Scenario 6 Base Run (Multicomponent Gas Jet)

Text Summary   ALOHA 5.2

```
SITE DATA INFORMATION:
Location: CONCORD, MASSACHUSETTS
Building Air Exchanges Per Hour: 0.50 (Enclosed office)
Time: July 10, 1995  1530 hours EDT (User specified)

CHEMICAL INFORMATION:
Chemical Name: PSEUDO GAS FOR SCENARIO 6
Molecular Weight: 31.25 kg/kmol
TLV-TWA: -unavail-              IDLH: -unavail-
Footprint Level of Concern: 50000 ppm
Boiling Point: -88.63° C
Vapor Pressure at Ambient Temperature: greater than 1 atm
Ambient Saturation Concentration: 1,000,000 ppm or 100.0%

ATMOSPHERIC INFORMATION: (MANUAL INPUT OF DATA)
Wind: 5 meters/sec from W at 10 meters
No Inversion Height
Stability Class: D            Air Temperature: 30° C
Relative Humidity: 75%        Ground Roughness: 20 centimeters
Cloud Cover: 5 tenths

SOURCE STRENGTH INFORMATION:
Direct Source: 5 kilograms/sec
Source Height: 2 meters
Release Duration: 60 minutes
Release Rate: 300 kilograms/min
Total Amount Released: 18,000 kilograms
Note: This chemical may flash boil and/or result in two phase flow.

FOOTPRINT INFORMATION:
Model Run: Heavy Gas
User-specified LOC: 50000 ppm
Max Threat Zone for LOC: 28 meters
Note: Footprint was not drawn because effects of
    near-field patchiness make dispersion predictions
    unreliable for short distances.

TIME DEPENDENT INFORMATION:
Concentration Estimates at the point:
Downwind:         1500 meters
Off Centerline: 0 meters
Max Concentration:
    Outdoor: 73.6 ppm
    Indoor:  27.5 ppm
Note: Indoor graph is shown with a dotted line.
```

APPENDIX G

Scenario 7: Transient (Mitigated) Area-Source Release of Hydrogen Fluoride

Scenario 7 is concerned with a release of hydrogen fluoride (HF) that is stored as a liquid at a temperature of 308 K and at a pressure of 4 atm (see Figure G-1). It is assumed that the liquid HF flashes when released, forming a two-phase aerosol cloud. The mass fraction of vapor in the cloud is 0.104 [calculated using equation (4-14)]. The cloud density is 7.98 kg m^{-3} [calculated using equation (5-16)], or about seven times denser than air. The aerosol cloud is assumed to initially spread out along the ground, forming an effective ground-based area source with diameter 24 m and temperature 292.7 K (the boiling point of HF). The mass emission rate from the area source is assumed to be 10 kg s^{-1} for the first one minute, and then is sharply reduced to 1 kg s^{-1} for the next nine minutes due to some mitigation measure. The source term is assumed to completely shut off after ten minutes. The ambient atmosphere is assumed to be stable (class F) with a wind speed of 2 m s^{-1} at a reference height of 10 m. Table 9-2 and Table G-1 provide the properties of HF, the source configuration, the meteorological data, and the site conditions for Scenario 7.

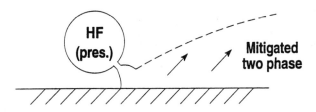

Figure G-1. Schematic diagram of Scenario 7, the short duration HF release.

TABLE G-1

Input Data Archive for Scenario 7, Short-Duration HF Release. (Data Taken from Table 9-2, Which Covers All Seven Scenarios)

HYDROGEN FLUORIDE PROPERTIES	
molecular weight	$M = 20.01$ kg kmole^{-1}
normal boiling point	$T_b = 292.7$ K
latent heat of evaporation	$H_{vap} = 373,000$ J kg^{-1}
specific heat at constant pressure for vapor	$c_{pv} = 2528$ J kg^{-1} K^{-1}
specific heat at constant pressure for liquid	$c_{pl} = 1450$ J kg^{-1} K^{-1}
density of liquid	$\rho_l = 987$ kg m^{-3}

SOURCE CONFIGURATION	
storage temperature	$T_s = 308$ K
release temperature	$T_{po} = 292.7$ K (normal boiling point)
storage pressure	$p_s = 10$ atm
storage phase	liquid
release phase	two-phase (liquid plus vapor)
initial cloud diameter	$D_o = 24$ m
flash fraction	$f = 0.104$
two-phase mixture density	$\rho_{po} = 7.98$ kg m^{-3}
source strength	$Q = 10$ kg s^{-1} for first 60 s $Q = 1$ kg s^{-1} for next 540 s

METEOROLOGICAL CONDITIONS	
temperature	$T_a = 293$ K at height of 10 m
wind speed	$u = 2$ m s^{-1} at height of 10 m
stability class	F
relative humidity	RH = 90%
Monin-Obukhov length	$L = 14.3$ m
friction velocity	$u* = 0.0986$ m s^{-1}
surface roughness	$z_0 = 0.1$ m (farmland)

SENSITIVITY RUNS	
RUN 1	Change relative humidity from 90% to 25%
RUN 2	Extend duration of initial 10 kg s^{-1} source term from 60 s to 300 s
RUN 3	Increase surface roughness, z_0, from 0.1 m to 0.5 m

CONCENTRATIONS OF INTEREST			
averaging time	$T_a = 600$ s	IDLH	$C = 30$ ppm

Before proceeding, it is advisable to calculate the initial Richardson number, Ri_0, in order to decide whether dense gas models should be applied. Equation (5-1), for continuous ground-level area sources, is valid for calculating Ri_0:

$$Ri_0 = \frac{g(\rho_{po} - \rho_a)V_{co}}{\rho_a u_*^3 D} \tag{G-1}$$

The following parameters are given in Table G-1 or can be calculated from known variables:

ρ_{po} = 7.98 kg m^{-3}

ρ_a = 1.20 kg m^{-3}

g = 9.8 m s^{-2}

u_* = 0.0986 m s^{-1}

D = 24 m

V_{co} = $Q/(\pi\rho_{po}) = 0.40$ m^3 s^{-1}

Therefore Ri_0 equals 957, which is a factor of 20 larger than the critical value of $Ri_0 = 50$, implying that it is appropriate to apply dense gas vapor cloud dispersion models to this scenario.

The DEGADIS and HGSYSTEM models have been applied to this scenario. HGSYSTEM directly treats HF. DEGADIS can treat HF only through a set of assumptions concerning the thermodynamics effects. Both models were applied in "transient" mode (i.e., assuming a time-varying source term), given the prescribed ground-level area source inputs. Model predictions are listed in Table G-2 and are plotted in Figures G-2 through G-5.

The concentrations predicted by the two models and plotted in Figure G-2 are in fair agreement for the first 500 m downwind, but then HGSYSTEM steadily predicts larger concentrations than DEGADIS. The difference is a factor of four at a distance of 2000 m. This causes a larger difference in the "penetration" distance of the IDLH concentration, 30 ppm, since DEGADIS predicts that the 30 ppm contour extends to a distance of about 2000 m, whereas HGSYSTEM predicts (by extrapolating the curve in Figure G-2) that the 30 ppm contour extends to a distance of about 10,000 m.

The predicted cloud half-widths (lateral distances to the 30 ppm contour) are listed in the right-hand columns in Table G-2 and are plotted in Figure G-4. In general, DEGADIS predicts a wider cloud than HGSYSTEM by about a factor of two at all downwind distances up to 1500 m, after which the DEGADIS 30 ppm contour quickly closes off.

TABLE G-2
Predicted Cloud Centerline Concentrations and Half-Widths (Lateral Distance to 30 ppm Contour) as a Function of Downwind Distance for Base Runs and Sensitivity Runs for DEGADIS and HGSYSTEM for Scenario 7 (Short-Duration HF Release)

		SCENARIO 7			
		Centerline Concentration (ppm)		Lateral Distance to 30 ppm (m)	
		DEGADIS	HGSYSTEM	DEGADIS	HGSYSTEM
Base Run:	100 m	9510	11924	163	68
RH=90%	200 m	2112	883	209	98
T_m=1min	500 m	363	395	260	111
z_0=0.1m	1000 m	96	229	264	138
	1500 m	44	152	223	206
	2000 m	25	108	n/a	226
Sensitivity Run 1:	100 m	9708	28543	168	55
RH=25%	200 m	2122	8152	209	81
	500 m	365	1837	260	128
	1000 m	97	438	265	160
	1500 m	44	257	224	172
	2000 m	25	164	n/a	176
Sensitivity Run 2:	100 m	68540	83886	94	123
T_m=5min	200 m	6527	9116	237	136
	500 m	789	587	359	202
	1000 m	188	365	404	256
	1500 m	81	258	394	283
	2000 m	45	191	342	298
Sensitivity Run 3:	100 m	3666	4488	156	62
z_0=0.5m	200 m	920	709	189	64
	500 m	190	352	221	116
	1000 m	57	172	203	149
	1500 m	27	110	n/a	189
	2000 m	16	75	n/a	208

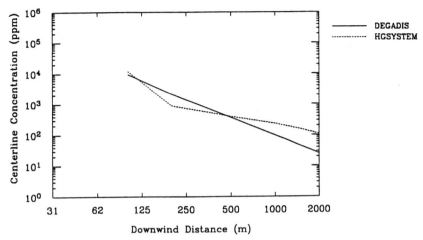

Figure G-2. Predicted cloud centerline concentration as a function of downwind distance for DEGADIS and HGSYSTEM for Scenario 7 (short-duration HF release).

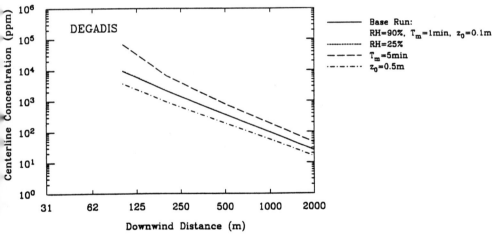

Figure A-3. Predicted cloud centerline concentration as a function of downwind distance for DEGADIS for the base run and three sensitivity runs for Scenario 7 (short-duration HF release).

The results of the three sensitivity runs with DEGADIS are listed in Table G-2 and are plotted in Figure G-3 (concentrations) and Figure G-5 (cloud half-widths). With the first sensitivity run (relative humidity decreased from 90% to 25%), predicted concentrations are relatively unchanged for DEGADIS, which is known to treat HF thermodynamics in a simplified manner. However, predicted concentrations increase by factors of 1.6 (at $x = 2000$ m) to 9 (at a distance of 200 m) for HGSYSTEM, which contains a complete description of HF thermodynamics. With lower relative humidity, less HF reactions occur and the cloud remains dense for a greater distance downwind.

The changes in predicted concentrations resulting from the two sensitivity runs with increased source duration time and increased surface roughness agree with expectations. As the duration of the 10 kg s^{-1} emission rate increases from 1 to 5 minutes, the predicted concentrations increase by

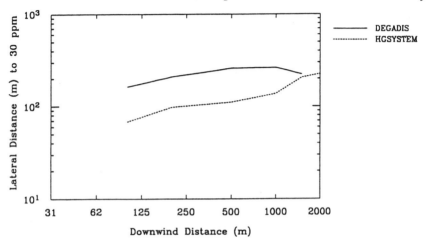

Figure G-4. Predicted cloud half-width (lateral distance to 30 ppm contour) as a function of downwind distance for DEGADIS and HGSYSTEM for Scenario 7 (short-duration HF release).

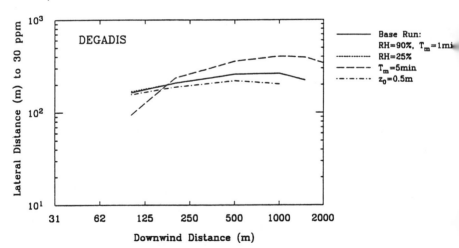

Figure G-5. Predicted cloud half-width (lateral distance to 30 ppm contour) as a function of downwind distance for DEGADIS for the base run and three sensitivity runs for Scenario 7 (short-duration HF release).

a factor of three to seven near the source ($x = 100$ to 200 m), and by a factor of two far from the source ($x = 1000$ to 2000 m). As the surface roughness increases by a factor of five, predicted concentrations decrease by a factor of two or three near the source and by about 30 or 40% far from the source, in agreement with the results of field observations and model sensitivity studies reported by Hanna et al. (1993).

The cloud half-widths plotted in Figure G-5 illustrate that there is little change for the sensitivity run with lower relative humidity. There is also little change in half-width when the source duration is increased from 1 to 5 minutes. When the roughness is increased from 0.1 to 0.5 m, the half-width near the source is reduced due to the inhibition of the vapor-blanket effect, but then is increased by about 50% at larger downwind distances due to the increase in ambient turbulence.

The diskette in the pocket inside the back cover contains the output files for all runs. A hard copy example of the DEGADIS output is given in Table G-3.

TABLE G-3 DEGADIS Output File for Scenario 7

```
0    TITLE BLOCK

TRIAL AND EXPERIMENT: S7A          Scenario 7
CHEMICAL: Hydrogen Fluoride
TRANSIENT SIMULATION IN ISOTHERMAL MODE
DENSITY EFFECTS OF AEROSOL SIMULATED

     Wind velocity at reference height        2.00    m/s
     Reference height                        10.00    m
0    Surface roughness length                 .100    m
0    Pasquill Stability class                 F
0    Monin-Obukhov length                    14.3     m
     Gaussian distribution constants
                 Specified averaging time   600.00    s
                              Delta          .06740
                              Betay          .90000
                              Alpha          .47672
0    Wind velocity power law constant         .08848   m/s
0    Friction velocity                        293.00   K
     Ambient Temperature                      1.000    atm
     Ambient Pressure                       1.321E-02  kg/kg BDA
     Ambient Absolute Humidity               90.00     %
     Ambient Relative Humidity
```

Input:	Mole fraction	CONCENTRATION OF C	GAS DENSITY
		kg/m**3	kg/m**3
	.00000	.00000	1.19648
	.01000	.00840	1.20346
	.02000	.01696	1.21073
	.03000	.02567	1.21814
	.04000	.03455	1.22569
	.05000	.04359	1.23337
	.10000	.09148	1.27408
	.15000	.14431	1.31900
	.20000	.20291	1.36882
	.25000	.26828	1.42439
	.30000	.34165	1.48676
	.35000	.42459	1.55727

```
          .40000          .51911          1.63763
          .45000          .62781          1.73004
          .50000          .75414          1.83744

          .55000          .90278          1.96379
          .60000         1.08019          2.11462
          .65000         1.29563          2.29777

          .70000         1.56281          2.52491
          .75000         1.90288          2.81401
          .80000         2.35041          3.19447

          .85000         2.96588          3.71770
          .90000         3.86564          4.48262
          .95000         5.30586          5.70700

         1.00000         7.98248          7.98248
```

0   Specified Gas Properties:

```
    Molecular weight:                                    20.010
    Release temperature:                                 293.00      K
    Density at release temperature and ambient pressure:  7.9825     kg/m**3
    Average heat capacity:                                .00000     J/kg K
    Upper mole fraction contour:                    3.00000E-05
    Lower mole fraction contour:                    1.00000E-07
    Height for isopleths:                                 .00000     m
```

Source input data points

   Initial (pure contaminant) mass in cloud: .00000   kg

Time	Contaminant Mass Rate	Source Radius	Contaminant Mass Fraction	Temperature	Enthalpy
s	kg/s	m	kg contam/kg mix	K	J/kg
.00000	10.000	12.000	1.0000	293.00	.00000
60.000	10.000	12.000	1.0000	293.00	.00000
61.000	1.0000	12.000	1.0000	293.00	.00000
540.00	1.0000	12.000	1.0000	293.00	.00000
541.00	.00000	.00000	1.0000	293.00	.00000
542.00	.00000	.00000	1.0000	293.00	.00000

TABLE G-3 (Continued)

- o Calculation procedure for ALPHA: 1
- o Entrainment prescription for PHI: 3
- o Layer thickness ratio used for average depth: 2.1500
- o Air entrainment coefficient used: .590
- o Gravity slumping velocity coefficient used: 1.150
- o Isothermal calculation
- o Heat transfer not included
- o Water transfer not included

***** CALCULATED SOURCE PARAMETERS *****

Time sec	Gas Radius m	Height m	Qstar kg/m**2/s	SZ(x*L/2.) m	Mole frac C	Density kg/m**3	Rich No.
.000000	12.0000	1.100000E-05	1.823204E-03	5.040302E-02	1.00000	7.98248	.756144
1.000000E-02	12.0004	3.640675E-05	1.823166E-03	5.040458E-02	.999996	7.98224	.756144
2.000000E-02	12.0010	6.181031E-05	1.823105E-03	5.040670E-02	.999991	7.98193	.756144
6.105012E-02	12.0047	1.660419E-04	1.822723E-03	5.041973E-02	.999962	7.98002	.756144
.102100	12.0099	2.701618E-04	1.822192E-03	5.043786E-02	.999921	7.97737	.756144
.296855	12.0474	7.618030E-04	1.818358E-03	5.056897E-02	.999627	7.95828	.756144
.491609	12.1001	1.248149E-03	1.813009E-03	5.075272E-02	.999214	7.93171	.756144
1.30135	12.4222	3.188121E-03	1.781138E-03	5.186894E-02	.996714	7.77469	.756144
2.11108	12.8554	4.963415E-03	1.740393E-03	5.335083E-02	.993417	7.57732	.756144
5.86110	15.5521	1.093698E-02	1.530312E-03	6.212201E-02	.974198	6.61768	.756144
9.61112	18.6715	1.416602E-02	1.352401E-03	7.135479E-02	.955362	5.88398	.756144
14.8536	23.1189	1.627227E-02	1.170346E-03	8.307901E-02	.932699	5.20859	.756144
20.0961	27.4585	1.700120E-02	1.041626E-03	9.311555E-02	.914697	4.78020	.756144
33.4161	37.7490	1.649522E-02	8.393665E-04	.112305	.884258	4.20691	.756144
46.7362	47.0655	1.488261E-02	7.224754E-04	.124968	.867818	3.95562	.756144
58.8071	54.8073	1.320165E-02	6.512400E-04	.132478	.860453	3.85352	.756144
66.3903	54.5616	1.366130E-02	6.539171E-04	.141731	.840169	3.60045	.756144
81.2293	51.9265	1.498488E-02	6.775418E-04	.163048	.791721	3.12390	.756144
100.191	47.9261	1.607783E-02	7.160725E-04	.190788	.729559	2.68723	.756144
111.769	44.9770	1.627288E-02	7.473878E-04	.206919	.693295	2.49173	.756144
126.953	40.3639	1.578717E-02	8.032516E-04	.224412	.651602	2.30465	.756144

142.136	34.7523	1.425895E-02	8.871258E-04	.232569	2.19607	.623410	.756144
149.805	31.5296	1.304385E-02	9.464815E-04	.230329	2.17798	.618377	.756144
157.473	28.0989	1.152241E-02	1.022349E-03	.221535	2.19721	.623723	.756144
163.376	25.3910	1.016591E-02	1.094658E-03	.209178	2.25038	.637925	.756144
169.278	22.7312	8.694390E-03	1.180210E-03	.191477	2.35561	.663752	.756144
174.477	20.5404	7.364352E-03	1.265157E-03	.171912	2.51103	.697573	.756144
179.676	18.6084	6.081328E-03	1.354505E-03	.150082	2.75701	.741043	.756144
185.825	16.7843	4.743009E-03	1.454997E-03	.124523	3.18850	.799261	.756144
191.973	15.5103	3.702697E-03	1.536666E-03	.102870	3.77425	.854427	.756144
197.848	14.7453	3.014310E-03	1.590716E-03	8.769802E-02	4.40743	.895897	.756144
203.724	14.1025	2.579478E-03	1.623587E-03	7.772652E-02	4.99563	.924170	.756144
209.319	14.0721	2.333073E-03	1.640983E-03	7.191129E-02	5.43424	.940948	.756144
214.913	13.9513	2.191001E-03	1.650089E-03	6.848567E-02	5.73695	.950917	.756144
218.216	13.9108	2.138426E-03	1.653116E-03	6.719871E-02	5.86093	.954481	.756144
221.520	13.8848	2.101443E-03	1.655038E-03	6.628403E-02	5.95292	.957365	.756144
226.275	13.8638	2.066942E-03	1.656559E-03	6.541973E-02	6.04311	.959911	.756144
235.323	13.8511	2.035822E-03	1.657407E-03	6.462076E-02	6.12987	.962285	.756144
239.615	14.0867	1.961883E-03	1.638420E-03	6.468284E-02	6.14927	.962806	.756144
253.832	13.9437	1.913358E-03	1.649643E-03	6.373620E-02	6.24126	.965229	.756144
279.990	13.8759	1.889696E-03	1.655027E-03	6.326440E-02	6.28868	.966449	.756144
540.356	13.8364	1.884227E-03	1.658248E-03	6.322272E-02	6.28890	.966455	.756144
542.226	12.7301	1.820793E-03	1.755762E-03	6.473065E-02	5.98764	.958355	.756144
545.227	10.4758	1.644485E-03	2.005016E-03	6.721260E-02	5.45554	.941188	.756144
547.090	8.84527	1.476087E-03	2.248006E-03	6.835294E-02	5.13181	.929712	.756144
548.119	7.83243	1.354490E-03	2.439694E-03	6.864829E-02	4.95213	.922329	.756144
549.147	6.70396	1.203578E-03	2.707790E-03	6.857240E-02	4.77010	.914230	.756144
549.838	5.85600	1.079754E-03	2.963612E-03	6.814628E-02	4.64459	.908240	.756144
550.528	4.90341	9.300705E-04	3.334978E-03	6.717664E-02	4.51496	.901670	.756144
550.954	4.24185	8.198124E-04	3.671022E-03	6.612298E-02	4.43114	.897196	.756144
551.380	3.49540	6.892418E-04	4.171077E-03	6.443102E-02	4.34318	.892297	.756144
551.635	2.98871	5.970859E-04	4.623129E-03	6.289468E-02	4.28683	.889041	.756144
551.890	2.41315	4.889524E-04	5.317463E-03	6.061943E-02	4.22692	.885475	.756144
552.036	2.03804	4.166331E-04	5.935623E-03	5.873678E-02	4.18964	.883199	.756144

TABLE G-3 (Continued)

552.182	1.60862	3.320413E-04	6.918331E-03	5.601908E-02	.880706	4.14964	.756144
552.261	1.34088	2.784039E-04	7.778363E-03	5.390952E-02	.879177	4.12553	.756144
552.340	1.03165	2.155801E-04	9.197615E-03	5.087118E-02	.877509	4.09956	.756144
552.380	.848627	1.779987E-04	1.041083E-02	4.864162E-02	.876529	4.08448	.756144
552.420	.635328	1.338049E-04	1.248974E-02	4.540145E-02	.875469	4.06830	.756144
552.430	.577733	1.218086E-04	1.325297E-02	4.436316E-02	.875175	4.06385	.756144
552.439	.530143	1.031353E-04	1.397961E-02	4.344299E-02	.874866	4.05916	.756144
552.461	.530143	1.206765E-05	1.397819E-02	4.347663E-02	.874630	4.05560	.756144
552.471	.530143	.000000	1.397819E-02	4.347663E-02	.875493	4.05560	.000000

o   Sorted values for each specified time.
o   X-Direction correction was applied.
    Coefficient: .17000
    Power: .97000
    Minimum Distance: 50.000 m

o Time after beginning of spill   255.0000   sec

Distance	Mole Fraction	Concentration	Density	Temperature	Half Width	Sz	Sy	Width at z= .00 m to:	
								1.000E-05mole%	3.000E-03mole%
(m)		(kg/m**3)	(kg/m**3)	(K)	(m)	(m)	(m)	(m)	(m)
14.6	.166	.163	1.3346	293.	29.7	.561	.788	32.7	32.1
19.4	.133	.127	1.3043	293.	38.8	.570	2.93	49.9	47.4
24.1	.110	.104	1.2849	293.	45.8	.598	4.67	63.3	59.3
28.9	.100	9.459E-02	1.2767	293.	49.9	.616	5.80	71.6	66.6
33.8	9.386E-02	8.823E-02	1.2713	293.	52.7	.629	6.60	77.3	71.6
38.7	8.663E-02	8.109E-02	1.2653	293.	54.5	.648	7.28	81.5	75.2
43.7	7.882E-02	7.337E-02	1.2587	293.	55.6	.669	7.92	84.9	78.0
48.7	6.989E-02	6.455E-02	1.2512	293.	56.1	.694	8.54	87.6	80.1
53.9	5.944E-02	5.427E-02	1.2425	293.	55.9	.725	9.16	89.4	81.3
59.1	4.686E-02	4.203E-02	1.2320	293.	54.2	.775	9.70	89.4	80.7
64.3	3.348E-02	2.932E-02	1.2212	293.	51.3	.836	10.2	87.6	78.3
69.7	2.154E-02	1.831E-02	1.2119	293.	47.4	.911	10.6	84.4	74.5

Distance (m)	Mole Fraction	Concentration (kg/m**3)	Density (kg/m**3)	Temperature (K)	Half Width (m)	Sz (m)	Sy (m)	Width at z= 1.000E-05molek (m)	.00 m to: 3.000E-03molek (m)
75.2	1.611E-02	1.364E-02	1.2079	293.	41.4	1.03	10.4	77.4	67.5
80.7	9.269E-03	7.788E-03	1.2030	293.	34.5	1.18	10.1	68.7	58.7
86.4	4.926E-03	4.123E-03	1.1999	293.	25.3	1.41	9.30	55.9	46.3
92.1	1.360E-03	1.134E-03	1.1974	293.	12.8	1.79	7.54	36.1	27.5

For the ULC of   3.00000E-03 mole percent, and the LLC of   1.00000E-05 mole percent:

The mass of contaminant between the ULC and LLC is: .00000   kg.
The mass of contaminant above the LLC is: 284.09   kg.

Time after beginning of spill   435.0000   sec

Distance (m)	Mole Fraction	Concentration (kg/m**3)	Density (kg/m**3)	Temperature (K)	Half Width (m)	Sz (m)	Sy (m)	Width at z= 1.000E-05molek (m)	.00 m to: 3.000E-03molek (m)
17.7	.237	.251	1.4093	293.	23.2	.437	2.83	34.1	31.8
22.3	.195	.197	1.3636	293.	35.8	.411	4.95	54.8	50.6
27.0	.155	.150	1.3241	293.	47.3	.428	6.95	73.7	67.8
31.9	.118	.110	1.2896	293.	55.9	.480	8.75	88.8	81.3
36.8	8.795E-02	7.999E-02	1.2643	293.	62.5	.558	10.4	101.	92.1
41.9	6.705E-02	5.986E-02	1.2472	293.	67.2	.659	12.0	111.	101.
46.8	4.530E-02	3.951E-02	1.2299	293.	71.9	.757	13.5	121.	109.
52.0	3.283E-02	2.829E-02	1.2204	293.	75.1	.891	15.0	129.	115.
57.2	2.493E-02	2.130E-02	1.2144	293.	77.9	1.01	16.3	136.	120.
62.5	2.041E-02	1.732E-02	1.2110	293.	80.3	1.09	17.2	141.	124.
67.9	1.848E-02	1.566E-02	1.2096	293.	81.7	1.15	17.9	144.	127.
73.4	1.652E-02	1.397E-02	1.2082	293.	82.1	1.22	18.4	146.	128.
79.0	1.474E-02	1.245E-02	1.2069	293.	81.6	1.28	18.8	146.	128.
84.6	1.294E-02	1.091E-02	1.2056	293.	80.3	1.36	19.1	146.	127.
90.3	1.107E-02	9.317E-03	1.2042	293.	77.8	1.45	19.3	144.	125.
96.0	9.186E-03	7.717E-03	1.2029	293.	73.6	1.56	19.3	139.	120.
102.	7.414E-03	6.218E-03	1.2016	293.	67.7	1.70	19.1	132.	113.
108.	5.717E-03	4.788E-03	1.2005	293.	60.4	1.86	18.7	122.	103.

TABLE G-3 (Continued)

For the ULC of 3.00000E-03 mole percent, and the LLC of 1.00000E-05 mole percent:

The mass of contaminant between the ULC and LLC is: .00000 kg.
The mass of contaminant above the ULC is: 353.68 kg.

Time after beginning of spill   615.0000 sec

Distance (m)	Mole Fraction	Concentration (kg/m**3)	Density (kg/m**3)	Temperature (K)	Half Width (m)	Sz (m)	Sy (m)	Width at z= .00 m to: 1.000E-05mole% (m)	3.000E-03mole% (m)
11.7	.131	.125	1.3027	293.	34.4	.262	4.99	53.3	49.0
16.2	.170	.171	1.3414	293.	37.0	.257	5.25	57.0	52.5
20.8	.179	.179	1.3482	293.	39.6	.264	5.60	61.0	56.3
25.5	.177	.177	1.3468	293.	42.1	.283	6.02	65.1	60.0
30.3	.130	.124	1.3014	293.	47.9	.356	7.38	75.8	69.5
35.1	9.058E-02	8.258E-02	1.2665	293.	54.6	.445	9.01	88.1	80.3
40.1	6.464E-02	5.723E-02	1.2450	293.	61.3	.531	10.6	100.	90.9
45.2	5.702E-02	5.052E-02	1.2393	293.	66.7	.633	12.2	111.	100.
50.4	4.617E-02	4.032E-02	1.2306	293.	73.1	.721	13.8	123.	111.
55.6	3.892E-02	3.368E-02	1.2249	293.	79.9	.807	15.5	136.	121.
60.8	3.411E-02	2.953E-02	1.2214	293.	86.3	.898	17.2	148.	132.
66.2	2.679E-02	2.286E-02	1.2158	293.	89.8	1.01	18.6	156.	138.
71.8	2.174E-02	1.847E-02	1.2120	293.	91.4	1.15	19.9	161.	142.
77.3	1.791E-02	1.516E-02	1.2092	293.	92.5	1.29	21.2	166.	146.
83.1	1.453E-02	1.227E-02	1.2067	293.	92.1	1.46	22.2	169.	148.
88.8	1.200E-02	1.011E-02	1.2049	293.	92.4	1.64	23.5	173.	150.
94.5	1.008E-02	8.473E-03	1.2035	293.	93.1	1.79	24.6	177.	152.
100.	8.676E-03	7.285E-03	1.2025	293.	94.1	1.90	25.4	180.	155.
114.	4.210E-03	3.521E-03	1.1994	293.	52.0	2.06	18.0	111.	92.0
120.	2.946E-03	2.461E-03	1.1985	293.	42.2	2.31	17.0	96.9	78.7
126.	1.833E-03	1.534E-03	1.1978	293.	30.2	2.64	15.4	78.4	61.4
132.	1.009E-03	8.418E-04	1.1972	293.	14.3	3.15	12.2	51.4	37.2

106.	7.655E-03	6.422E-03	1.2018	293.	94.9	1.98	26.0	182.	156.
112.	6.842E-03	5.735E-03	1.2012	293.	94.6	2.06	26.5	183.	156.
118.	6.092E-03	5.104E-03	1.2007	293.	93.5	2.15	26.8	183.	155.
124.	5.354E-03	4.483E-03	1.2002	293.	91.5	2.25	27.0	181.	153.
130.	4.619E-03	3.865E-03	1.1997	293.	88.2	2.37	27.1	177.	149.
136.	3.893E-03	3.255E-03	1.1992	293.	82.9	2.51	26.9	170.	142.
143.	3.192E-03	2.667E-03	1.1987	293.	75.9	2.69	26.5	161.	133.
149.	2.507E-03	2.093E-03	1.1982	293.	67.5	2.89	25.7	149.	122.
156.	1.894E-03	1.581E-03	1.1978	293.	57.7	3.14	24.7	135.	108.
162.	1.388E-03	1.158E-03	1.1974	293.	46.5	3.45	23.2	118.	91.9
169.	9.453E-04	7.883E-04	1.1971	293.	32.7	3.84	20.8	95.7	71.4
175.	6.026E-04	5.024E-04	1.1969	293.	14.8	4.40	16.4	63.3	43.3

For the ULC of 3.00000E-03 mole percent, and the LLC of 1.00000E-05 mole percent:

The mass of contaminant between the ULC and LLC is: .00000 kg.
The mass of contaminant above the LLC is: 464.38 kg.

References

AIChE/CCPS, 1989: *Guidelines for Chemical Process Quantitative Risk Analysis*. AIChE, 345 East 47th Street, New York, 585 pp.

AIChE/CCPS, 1995: Improved Aerosol Jet Model with Rainout, Absorption, and Reevaporation, Report by DNV Technica, Inc., Columbus, OH, available from AIChE, 345 East 47th Street, New York.

AIHA, 1995: American Industrial Hygiene Association. Emergency Response Planning Guidelines. AIHA, P.O. Box 27632, Richmond, VA 23261–7632.

Bell, R.P., 1978: Isopleth calculations for ruptures in sour gas pipelines. Energy Processing/Canada, 36–39.

Bird, P.B., W.E. Stewart, and E.N. Lightfoot, 1960: *Transport Phenomena*. John Wiley & Sons, 780 pp.

Blewitt, D.N., J.F. Yohn, R.P. Koopman, and T.C. Brown, 1987: Conduct of Anhydrous Hydrofluoric Acid Spill Experiments. *Proc. Int. Conf. on Vapor Cloud Modeling*, AIChE, New York, pp. 1–38.

Bloom, S.G., R.A. Just, and W.R. Williams, 1989: A Computer Program Simulating the Atmospheric Dispersion of UF_6 and other Reactive Gases having Positive, Neutral, or Negative Buoyancy. K/D-5694, MMES, ORGDP, Oak Ridge, TN 57831.

Bodurtha, F.T., 1961: The behavior of dense stack gases. *J. Air Poll. Control Assoc.*, 11, 431–435.

Briggs, G.A., 1984: Plume Rise and Buoyancy Effects. *Atmospheric Science and Power Production*, DOE/TIC-27601, 327–366.

Briggs, G.A., 1995: Field-measured dense gas plume characteristics and some parameterizations. *Proc. Int. Conf. on Modeling and Mitigating the Consequences of Accidental Releases of Hazardous Materials* (held in New Orleans), AIChE/CCPS, New York.

Brighton, P.W.M., 1989: The Effects of Natural and Man-Made Obstacles on Heavy Gas Dispersion. Safety and Reliability Directorate. UK Atomic Energy Authority. Wigshaw Lane, Culceth, Warrington, WA3 4NE, UK.

Brighton, P.W.M., 1990: Further verification of a theory for mass and heat transfer from evaporating pools. *J. Haz. Mat.*, 23, 215–234.

Brighton, P., A. Byrne, R. Cleaver, P. Courtiade, B. Crabol, R. Fitzpatrick, A. Girard, S. Jones, V. Lhomme, A. Mercer, D. Nedelka, C. Proux, and D. Webber, 1994: Comparison of heavy gas dispersion models for instantaneous releases. *J. Haz. Mat.*, 36, 193–208.

Brighton, P.W., A.J. Prince and D.M. Webber, 1985: Determination of cloud area and path from visual and concentration records. *J. Haz. Mat.*, **11**, 155–178.

Britter, R.E., 1995: Dispersion of two phase flashing releases - FLADIS field experiment. A further note on modelling flashing releases. Cambridge Environmental Research Consultants Ltd., Report FM89/3, for Commission of European Communities DGXII.

Britter, R.E. and W.H. Snyder, 1988: Fluid modeling of dense gas dispersion over a ramp. *J. Haz. Mat.*, **18**, 37–67.

Britter, R., 1980: The ground-level extent of a negatively buoyant plume in a turbulent boundary layer. *Atmos. Environ.*, **14**, 779–785.

Britter, R.E. and J. McQuaid, 1988: Workbook on the Dispersion of Dense Gases. HSE Report No. 17/1988, Health and Safety Executive, Sheffield, UK, 158 pages.

Britter, R.E., 1989: Atmospheric dispersion of dense gases. *Ann. Rev. Fluid Mech.*, **21**, 317–344.

Cambridge Environmental Research Consultants, 1990: GASTAR Dense Gas Dispersion Model Version 2.2 Users Guide. CERC, Cambridge, U.K.

Carslaw, H.S. and J.C. Jaeger, 1959: *The Conduction of Heat in Solids*. 2nd Ed., Oxford University Press, New York.

Cavanaugh, T.A., J.H. Siegell, and K.W. Steinberg, 1994: Simulation of vapor emissions from liquid spills. *J. Haz. Mat.*, **38**, 41–63.

Chan, S.T., D.L. Ermak, and L.K. Morris, 1987: FEM3 model simulations of selected Thorney Island Phase I trials. *J. Haz. Mat.*, **16**, 267–292.

Chan, S.T., 1994: FEM3C - An Improved Three-Dimensional Heavy-Gas Dispersion Model: Users Manual. UCRL-MA-116567. Rev. 1, Lawrence Livermore National Laboratory, Livermore, CA, 94551.

Chatwin, P.C., 1982: The use of statistics in describing and predicting the effects of dispersing gas clouds. *Dense Gas Dispersion*. (ed. by Britter and Griffiths), Elsevier, New York, 213–230.

Cox, W.M. and J.A. Tikvart, 1990: A statistical procedure for determining the best perform-ing air quality simulation model. *Atmos. Environ.*, **24A**, 2387–2395.

Daggupaty, S.M., 1988: Response to Accidental Release of Toxic Chemicals into the Atmosphere Using AQPAC. M.I. El-Sabh and T.S. Murty (eds.) *Natural and Man-Made Hazards.*, D. Reidel, 599–608.

DeNevers, N. 1984: Spread and downslope flow of negatively buoyant clouds. *Atmos. Environ.*, **18**, 2029–2032.

Deaves, D.M., 1984: Application of advanced turbulence models in determining the struc-ture and dispersion of heavy gas clouds. *Atmos. Disp. of Heavy Gases and Small Particles*. IUTAM Symp. (eds. G. Ooms, H. Tennekes) Springer-Verlag.

DeVaull, G.E., J.A. King, R.J. Lantzy, and D.J. Fontaine, 1995: Understanding Atmospheric Dispersion of Accidental Releases. AIChE/CCPS, 345 East 47th Street, New York, NY 10017.

DNV Technica, 1993: PHAST 4.2 Release Notes, App. I. DNV Technica, Inc., Temecula, CA.

Drivas, P.J., 1982: Calculation of evaporative emissions from multicomponent liquid spills. *Environ. Sci. Tech.*, **16**, 726–728.

Drivas, P.J., J.S. Sabnis and L.H. Teuscher, 1983: Model simulates pipeline, storage tank failures. *Oil and Gas Journal*, Sept. 12, 1983, 162–169.

DuPont, 1989: User's Manual for TRACE II. E.I. DuPont DeNemours, 5700 Corsa Avenue, Westlake Village, CA 91362.

Efron, B., 1987: Better bootstrap confidence intervals. *J. Am. Stat. Assoc.* **82**, 171–185.

Eidsvik, K.J., 1980: A model for heavy gas dispersion in the atmosphere. *Atmos. Environ.*, **14**, 769–777.

England, W.G., L.H. Teuscher, L.E. Hauser and B. Freeman, 1978: Atmospheric dispersion of liquefied natural gas vapor clouds using SIGMET, a three-dimensional time-dependent hydrodynamic computer model. 1978 Heat Transfer and Fluid Mechanics Institute.

EPA, 1990: OAQPS Control Cost Manual. Fourth Ed., EPA 450/3–90–006. United States Environmental Protection Agency, Office of Air Quality Planning and Standards.

EPA, 1994: Users Guide for TSCREEN - A Model for Screening Toxic Air Pollutant Concentrations (revised), Report EPA-454/B-94–023, USEPA, Research Triangle Park, NC 27711.

EPA, 1995: Compilation of Air Pollutant Emission Factors, Volume I: Stationary point and area sources, AP-42, Fifth Edition. U.S. Environmental Protection Agency, Office of Air Quality Planning and Standards, Research Triangle Park, NC 27711.

EPA, 1995: User's Guide for the Industrial Source Complex (ISC3) Dispersion Models, Volume II - Description of Model Algorithms. United States Environmental Protection Agency, Office of Air Quality Planning and Standards. EPA 454/B-95–003b, USEPA, RTP, NC 27711.

Ermak, D.L., 1990: User's Manual for SLAB: An Atmospheric Dispersion Model for Denser-than-Air Releases. UCRL-MA-105607, Lawrence Livermore National Laboratory, Livermore, CA.

Fauske, H.K., 1985: Flashing flows or: Some practical guidelines for emergency releases. *Plant/Operations Progress*, **4**, 132–134.

Fauske, H.K. and M. Epstein, 1988: Source term considerations in connection with chemical accidents and vapour cloud modelling. *J. Loss Prevention in the Process Industries*, **1**, 75–83.

Fay, J.A., 1986: The dispersion of dense gases in the atmosphere. *Proc. AMS Short Course on Atmos. Disp.*

Fisher, H.G., H.S. Forrest, S.S. Grossel, J.E. Huff, A.R. Muller, J.A. Noronha, D.A. Shaw, and B.J. Tilley, 1992: Emergency relief system design using DIERS technology: The Design Institute for Emergency Relief Systems (DIERS) project manual. Design Institute for Emergency Relief Systems, American Institute of Chemical Engineers, New York.

Fleischer, M.T., 1980: SPILLS, An evaporation/air dispersion model for chemical spills onland. Shell Devel. Center, Westhollow Res. Center, P.O. Box 1380, Houston, TX 77001.

Fryer, L.S. and G.D. Kaiser, 1979: DENZ - A computer program for the calculation of the dispersion of dense toxic or explosive gases in the atmosphere. SRD R 152 UKAEA, Culcheth, UK.

Gifford, F.A., 1976: Turbulence diffusion typing schemes - a review. *Nuc. Safety*, **17**, 68–86.

Golder, D., 1972: Relations between stability parameters in the surface layer. *Bound. Layer Meteorol.*, **3**, 46–68.

Goldwire, H.C., Jr., H.C. Rodean, R.T. Cederwall, E.J. Kansa, R.P. Koopman, J.W. McClure, T.G. McRae, L.K. Morris, L. Kamppinen, R.D. Kiefer, P.A. Urtiew, and C.D. Lind, 1983: Coyote series data report, LLNL/NWC 1981 LNG Spill Tests: Dispersion, vapor burn, and rapid phase transition. UCID-199953, Lawrence Livermore National Laboratory, Livermore, CA.

Goldwire, H.C., Jr., T.G. Mcrae, G.W. Johnson, D.L. Hipple, R.P. Koopman, J.W. McClure, L.K. Morris, and R.T. Cederwall, 1985: Desert Tortoise series data report: 1983 pressurized ammonia spills. UCID-20562, Lawrence Livermore National Laboratory, Livermore, CA.

Hanna, S.R., G.A. Briggs, and R.F. Hosker, Jr., 1982: *Handbook on Atmospheric Diffusion.* DOE/TIC-11223, Technical Information Center, U.S. Department of Energy, Oak Ridge, TN.

Hanna, S.R., 1984: The exponential pdf and concentration fluctuations in smoke plumes. *Bound. Lay. Meteorol.,* **29,** 361–376.

Hanna, S.R., 1989: Confidence limits for air quality model evaluations, as estimated by bootstap and jackknife resampling methods. *Atmos. Environ.,* **23,** 1385–1398.

Hanna, S.R. and J.C. Chang, 1992: Boundary-layer parameterizations for applied dispersion modeling over urban areas. *Bound. Lay. Meteorol.,* **58,** 229–259.

Hanna, S.R., 1992: Effects of surface roughness on short range dispersion. *Proc. Tenth Symposium on Turbulence and Diffusion,* AMS, Boston, MA, 102–105.

Hanna, S.R., J.C. Chang and D.G. Strimaitis, 1993: Hazardous gas model evaluation with field observations. *Atmos. Environ.,* **27A,** 2265–2285.

Hanna, S.R. and P.J. Drivas, 1993: Modeling VOC emissions and air concentrations from the Exxon Valdez oil spill. *J. Air & Waste Management Assoc.,* **43,** 298–309.

Havens, J.A. and T.O. Spicer, 1985: Development of an atmospheric dispersion model for heavier-than-air gas mixtures. Final report to the U.S. Coast Guard, CG-D-23–85. USCG HQ, Washington, DC.

Havens, J.A., 1988: A dispersion model for elevated dense gas jet chemical releases. Volumes I and II. EPA-450/4–88–006, U.S. Environmental Protection Agency.

Heidorn, K.C., M.C. Murphy, P.A. Irwin, H. Sahota, P.K. Misra, and R. Bloxan, 1992: Effects of obstacles on the spread of a heavy gas — Wind tunnel simulations. *J. Haz. Mat.,* **30,** 151–194.

Heinold, D., R. Paine, K. Walker and D. Smith, 1986: Evaluation of the AIRTOX dispersion algorithms using data from heavy gas field experiments. *Fifth Joint Conf. of Applications of Air PollutioMeteorology,* Research Triangle Park, NC, American Meteorological Society.

Hoot, T.G., R.N. Meroney and J.A. Peterka, 1973: Wind Tunnel Tests of Negatively Buoyant Plumes. CER73–74TGH-RNM-JAP-13, Colorado State Univ., Fort Collins, CO

Jones, S.J., A. Mercer, G.A.Tickle, D.M. Webber, and T. Wren, 1993: Initial Verification and Validation of DRIFT. UKAEA Report SRD/HSE R580.

Kantha, H.L., O.M. Phillips, and R.S. Azad, 1977: On Turbulent Entrainment at a Stable Density Interface. *J. Fluid Mech.* **79,** 753–768.

Kato, H. and O.M. Phillips, 1969: On the penetration of a turbulent layer into stratified fluid. *J. Fluid Mech.,* **37,** 643–655.

Kawamura, P. and D. Mackay, 1987: The evaporation of volatile liquids. *J. Haz. Mat.,* **15,** 343–364.

Konig, G., 1987: Windkanalmodellierung der Ausbreitung Stofallartig Freigesetzter Gase Schwerer als Luft. *Hanburger Geophysikalische Einzelschriften,* Series A, No. **85.**

Koopman, R.P., J. Baker, R.T. Cederwall, H.C. Goldwin, Jr., W.J. Hogan, L.M. Kamppinen, R.D. Kiefer, J.W. McClure, T.G. McRae, D.L. Morgan, L.K. Morris, M.W. Spann, Jr., and C.D. Lind, 1982: Burro Series Data Report. LLNL/NWC 1980 LNG Spill Tests, UICD-19075, Lawrence Livermore National Laboratory, Livermore, CA.

Kranenburg, C., 1984: Wind-induced entrainment in a stably-stratified fluid. *J. Fluid Mech.*, **145**, 253–273.

Kukkonen, J. and J. Nikmo, 1992: Modeling heavy gas cloud transport in sloping terrain. *J. Haz. Mat.*, **31**, 155–176.

Leonelli, P., C. Stramigioli, and G. Spadoni, 1994: The modeling of pool vaporization. *J. Loss Prev. Process Inc.*, **7**, 443–450.

Leung, J.C., 1986: A generalized correlation for one-component homogeneous equilibrium flashing choked flow. *AIChE Journal*, **32**, 1743–1746.

Leung, J.C. and M.A. Grolmes, 1988: A generalized correlation for flashing choked flow of initially subcooled liquid. *AIChE Journal*, **34**, 688–691.

Lofquist, K., 1954: Flow and Stress Near an Interface Between Stratified Liquids. *Physics of Fluids*, **3**.

Mackay, D. and R.S. Matsugu, 1973: Evaporation rates of liquid hydrocarbon spills on land and water. *Can. J. Chem Eng.*, **51**, 434–439.

Marotzke, K., 1988: Wind Tunnel Modeling of Density Current Interaction with Surface Obstacles. Meeting of CEC Project BA at TNO in Apeldoorn, The Netherlands, Sept. 29–30.

McFarlane, K., A. Prothero, J.S. Puttock, P.T. Roberts and H.W.M. Witlox, 1990: Development and Validation of Atmospheric Dispersion Models for Ideal Gases and Hydrogen Fluoride. Part I: Technical Reference Manual. Report No. TNER. 90.015, Shell Research Ltd., Thornton Research Centre, P.O. Box 1, Chester, U.K.

McQuaid, J., 1976: Unpublished data produced by McQuaid and supplied by R.E. Britter, produced for the report "Some Experiments on the Structure of Stably Stratified Shear Flows." (Tech Paper P21, SMRE, Sheffield, UK, 1976).

McQuaid, J., 1985: Objectives and design of the Phase I heavy gas dispersion trials. *J. Haz. Mat.*, **11**, 1–34.

McQuaid, J and B. Roebuck, 1985: Large Scale Field Trials on Dense Vapour Dispersion. Safety Engineering Laboratory, Research and Laboratory Services Division, Broad Lane, Sheffield, U.K.

Meroney, R.N., 1982: Wind-tunnel experiments on dense gas dispersion. *J. Haz. Mat.*, **9**, 85–106.

Meroney, R.N., 1987: Guidelines for fluid modeling of dense gas cloud dispersion. *J. Haz. Mat.*, **17**, 23–46.

Mills, M., 1988: Techniques for modeling jet releases for emergency prevention and preparedness applications. *1988 Hazardous Materials Spills Conf.*, Co-sponsored by the American Institute of Chemical Engineers and the National Response Team, Chicago, IL., 16–19 May.

NFPA, 1994: Fire Protection Guide to Hazardous Materials. 11th Edition. National Fire Protection Association, One Batterymarch Park, Quincy, MA 02269.

NIOSH, 1995: *NIOSH Pocket Guide to Chemical Hazards.* U.S. Department of Health and Human Services, Public Health Service, Centers for Disease Control and Prevention, National Institute for Occupational Safety and Health.

NOAA/HMRAD and EPA/CEPPO, 1992: ALOHA User's Manual and Theoretical Description. Reports available from NOAA/HMRAD, 7600 Sand Point Way N.E., Seattle, WA 98115.

Norris, H.L., 1995: Models and data on single phase and multiphase release rates from vessels and pipelines. Proceedings, Internat. Conf. and Workshop on Modeling and

Mitigating the Consequences of Accidental Releases of Hazardous Materials. AIChE, New York, 167–187.

Oke, T.R., 1978: *Boundary Layer Climates*. John Wiley and Sons, New York, 372 pp.

Ooms, G., A.P. Mahieu and F. Zelis, 1974: The plume path of vent gases heavier than air. First Int. Symp. on Loss Prevention and Safety Promotion in the Process Industries. The Hague.

Ooms, G. and N.J. Duijm, 1984: Dispersion of a stack plume heavier than air. Atmos. Disp. of Heavy Gases and Small Particles, IUTAM Symp. (eds, G. Ooms, H. Tennekes), Springer-Verlag, 1–23.

Panofsky, H.A. and J.A. Dutton, 1984: *Atmospheric Turbulence*. Wiley, New York, 397 pp.

Perry, R.H., D.W. Green, and J.O. Maloney, Ed., 1984: *Perry's Chemical Engineers' Handbook*. McGraw-Hill Book Company, 2336 pp.

Pleim, J., A. Venkatram and R. Yamartino, 1984: ADOM/TADAP Model Development Program. Vol. 4, The Dry Deposition Module. Ontario Ministry of the Environment, Rexdale, Ontario, Canada.

Puttock, J.S., D.R. Blackmore, G.W. Colenbrander, P.T. Davis, A. Evans, J.B. Homer, J.J. Redfearn, W.C. Van't Sant, and R.P. Wilson, 1984: Spill Tests of LNG and Refrigerated Liquid Propane on the Sea, Maplin Sands, 1980: Experimental Details of the Dispersion Tests. TNER.84.046, Shell Research Ltd., Thornton Research Centre, Combustion Division, P.O. Box 1, Chester, U.K.

Quest Consultants, Inc., 1990: FOCUS Input and Reference Manual. Quest Consultants Inc., P.O. Box 721387, Norman, OK 73070.

Radian Corporation, 1991: CHARM Emergency Response System Documentation. Radian Corporation, 8501 MoPac Blvd., Austin, TX 78720.

Raj, P.K., 1985: Summary of heavy gas spills modeling research. *Proc. of Heavy Gas (LNG/LPG) Workshop*. Toronto, 51–75.

Ramsdell, J.V., Jr., C.A. Simonen and K.W. Burk, 1993: Regional Atmospheric Transport Code for Hanford Emission Tracking (RATCHET). CDC Cont. No. 200-92-0503(CDC)/18620(BNW), Battelle, PNL, Richland, WA 99352.

Raphael, J.M., 1962: Prediction of temperature in rivers and reservoirs. *Proc. Amer. Soc. Civ. Eng., J. Power Division*, **88**, 157–165.

Riou, Y.J. and A.E. Saab, 1985: A three-dimensional numerical model for the dispersion of heavy gases over complex terrain. Paper VI.7, *Proc. of 15th Int. Tech. Meeting on Air Poll. Modeling and its Applic.*, NATO/CCMS.

Sallet, D.W., 1990: Critical two-phase mass flow rates of liquified gases. *J. Loss Prevention in the Process Industries*, **3**, 38–42.

Schatzmann, M., 1995: Accidental releases of heavy gases in urban areas. *Wind Climate in Cities* (J.E. Cermak et al., eds.), Elsevier Academic Publishers, The Netherlands, pp. 555–574.

Schulman, L.L., S.R. Hanna, and R. Britter, 1990: Effects of Structures on Toxic Vapor Dispersion. ESL-TR-90-99, ESL, AFESC, Tyndall AFB, FL 32403.

Sehmel, G.A., 1984: Deposition and Resuspension. Ch. 12 in *Atmospheric Science and Power Production*, DIE/TIC-27601, DE84005177, NTIC, USDOC, Springfield, VA 22161, pp 533–583.

Shaw, P. and F. Briscoe, 1978: Evaporation from Spills of Hazardous Liquids on Land and Water. SRD R 100 UKAEA, Culcheth, UK.

Slade, D.H., 1968: Diffusion from Instantaneous Sources. *Meteorology and Atomic Energy*, TID-24190, USAEC, 163–175.

Slinn, W.G.N., 1984: Precipitation Scavenging. Ch. 11 in *Atmospheric Science and Power Production*, DOE/TIC-27601, DE84005177, NTIC, USDOC, Springfield, VA 22161, pp 466–532

Spicer, T.O., 1985: Mathematical Modeling and Experimental Investigation of Heavier-than-Air Gas Dispersion in the Atmosphere. Ph.D. Dissertation, University of Arkansas, Fayetteville.

Spicer, T.V. and J.A. Havens, 1989: Users Guide for the DEGADIS 2.1 Dense Gas Dispersion Model. EPA-450/4/89–019, EPA/OAQPS, RTP, NC.

Spicer, T., 1990: Implementation of DEGADIS 2.1 on a Personal Computer. Prepared for American Petroleum Institute, 1220 L Street, Northwest, Washington, DC 20005.

Studer, D.W., B.A. Cooper, and L.C. Doelp, 1988: Vaporization and dispersion modeling of contained refrigerated liquid spills. *Plant/Operations Progress*, 7, 127–135.

Sutton, O.G., 1953: *Micrometeorology*. McGraw-Hill, New York.

Sykes, R.I., S.F. Parker, D.S. Henn, and W.S. Lewellen, 1993: Numerical simulation of ANATEX tracer data using a turbulence closure model for long-range dispersion. *J. Applied Meteorol.*, 33, 929–947.

Technica Ltd, 1989: PHAST Version 2.0 Theory Manual. Technica, London.

Touma, J.S., W.M. Cox, H. Thistle, and I. G. Zapert, 1995a: Performance evaluation of dense gas dispersion models. *J. Appl. Meteorol.*, 34, 603–615.

Touma, J.S. J.S. Irwin, J.A. Tikvart, and C.T. Coulter, 1995b: A review of procedures for updating air quality modeling techniques for regulatory programs. *J. Applied Meteorol.*, 34, 731–737.

Turner, D.B., 1994: *Workbook of Atmospheric Dispersion Estimates*. Second Edition. Lewis Publishers, 200 Corporate Blvd NW, Boca Raton, FL 33431–9868, 192 pp.

Urben, P.G., (ed.), 1994: Bretherick's Handbook of Reactive Chemical Hazards, Fifth Edition, Two Volumes. Butterworth-Heinemann.

van Ulden, A.P., 1974: On the spreading of a heavy gas released near the ground. *Proc. Loss Prevention and Safety Promotion in the Process Industries*, 221–226.

VDI, 1990: Dispersion of Heavy Gas Emissions by Accidental Releases - Safety Study. VDI 3783. Verein Dutscher Ingeniere, Germany.

Verschueren, K., 1983: *Handbook of Environmental Data on Organic Chemicals, Second Ed.*, Van Nostrand Reinhold, New York, 1310 pp.

Webber, D.M., S.J. Jones, G.A. Tickle, and T. Wren, 1992: A Model of a Dispersing Gas Cloud, and the Computer Implementation. I Near Instantaneous Release, II Steady Continuous Releases. UKAEA Reports SRD/HSE R586 (for part I) and R. 587 (for part II).

Webber, D.M. and J.S. Kukkonen, 1990: Modelling two-phase jets for hazard analysis. *J. Haz. Mat.*, 23, 167–182.

Wilson, D.J., 1979: The release and dispersion of gas from pipeline ruptures. Prepared for Alberta Environment.

Wilson, D.J., 1981: Along-wind diffusion of source transients. *Atmos. Environ.*, 15, 489–495.

Wilson, D.J., 1995: Concentration Fluctuations and Averaging Times in Vapor Clouds. CCPS Project No. 73, CCPS/AIChE, 345 East 47 Street, New York, NY 10017, 181 pages.

Wilson, D.J., and R.E. Britter, 1982: Estimates of building surface concentrations from nearby point sources. *Atmos. Environ.*, 16, 2631–2646.

Witlox, H.W.M., 1994a: The HEGADAS model for ground-level heavy-gas dispersion—I. Steady-state model. *Atmos. Environ.*, **28**, 2917–2932.

Witlox, H.W.M., 1994b: The HEGADAS model for ground-level heavy-gas dispersion—II. Time-dependent model. *Atmos. Environ.*, **28**, 2933–2946.

Witlox, H.W.M. and K. McFarlane, 1994: Interfacing dispersion models in the HGSYSTEM hazard-assessment package. *Atmos. Environ.*, **28**, 2947–2962.

Witlox, H.W.M., K. McFarlane, F.J. Rees and J.S. Puttock, 1990: Development and Validation of Atmospheric Dispersion Models for Ideal Gases and Hydrogen Fluoride. Part II: HGSYSTEM Program User's Manual. Report No. TNER. 90.016, Shell Research Limited, Thornton Research Center, P.O. Box 1, Chester, U.K.

Woodward, J.L., 1993: Expansion zone modeling of two-phase and gas discharges. *J. Haz. Mat.*, **33**, 307–318.

Woodward, J.L., 1995: Aerosol drop size correlation and corrected rainout data using models of drop evaporation, pool absorption, and pool evaporation. Int. Conf. and Workshop on Modeling and Mitigating the Consequencies of Accidental Releases of Hazardous Materials. AIChE/CCPS, 345 East 47th St., New York, NY 10017, 117–148.

Wu, J.M. and J.M. Schroy, 1979: Emissions from Spills. *APCA Spec. Conf. on Control of Specific (Toxic) Pollutants*, Pittsburgh.

Xiao-Yun, L., H. Leijdens and G. Ooms, 1986: An experimental verification of a theoretical model for the dispersion of a stack plume heavier than air. *Atmos. Environ.*, **20**, 1087–1094.

Index